It is often said that energy is the lifeblood of modern societies. It follows that if we want a society that is just and fair, then energy decisions must support these goals. Benjamin K. Sovacool's perceptive and innovative new book examines energy decision-making in this context. He makes a powerful case for energy justice.

James Gustave Speth, former dean of the Yale School of Forestry and Environmental Studies, founder and president of the World Resources Institute, and author of *America the Possible: Manifesto for a New Economy*

Benjamin K. Sovacool demonstrates through important case studies and original research the essential link between access to resources and individual and community opportunity and freedom. As we move into the twenty-first century, where population, environmental quality, and resource pressures will imprint themselves centrally on every aspect of society, this book provides a thought-provoking guide to the challenges ahead.

Daniel M. Kammen, Class of 1935 Distinguished Professor of Energy, University of California, Berkeley

Benjamin K. Sovacool, one of the smartest energy analysts anywhere, clearly explains that energy policy has as much to do with justice and ethics as it does economics and technology. Energy and Ethics fills a big hole in the policy debate moving us closer to a consensus beyond political stalemate. Essential reading.

David W. Orr, Paul Sears Distinguished Professor, Oberlin College

When we think of energy, we conjure up images of infrastructure such as coal mines, well fields, pipelines, power and plants. We worry about supply, demand, security, risk, public acceptance, costs, and even war. What we rarely think about is the twinned topics of ethics and justice. With this new book, Benjamin K. Sovacool has changed all that. Energy studies will never be the same. And that's a good thing. In these times of increasing competition for the energy resources that await discovery beneath mountains, prairies, deserts, forests, permafrost, oceans, and Arctic ice, what could be more important than justice?

Martin J. (Mike) Pasqualetti, Senior Sustainability Scientist, Global Institute of Sustainability, and Professor of Geographical Sciences and Urban Planning, Arizona State University

Energy, Climate and the Environment

Series Editor: **David Elliott**, Emeritus Professor of Technology, Open University, UK

Titles include:

Luca Anceschi and Jonathan Symons (*editors*)
ENERGY SECURITY IN THE ERA OF CLIMATE CHANGE
The Asia-Pacific Experience

Philip Andrews-Speed
THE GOVERNANCE OF ENERGY IN CHINA
Implications for Future Sustainability

Ian Bailey and Hugh Compston (*editors*)
FEELING THE HEAT
The Politics of Climate Policy in Rapidly Industrializing Countries

Mehmet Efe Biresselioglu
EUROPEAN ENERGY SECURITY
Turkey's Future Role and Impact

David Elliott (*editor*)
NUCLEAR OR NOT?
Does Nuclear Power Have a Place in a Sustainable Future?

David Elliott (*editor*)
SUSTAINABLE ENERGY
Opportunities and Limitations

Horace Herring and Steve Sorrell (*editors*)
ENERGY EFFICIENCY AND SUSTAINABLE CONSUMPTION
The Rebound Effect

Horace Herring (*editor*)
LIVING IN A LOW-CARBON SOCIETY IN 2050

Matti Kojo and Tapio Litmanen (*editors*)
THE RENEWAL OF NUCLEAR POWER IN FINLAND

Antonio Marquina (*editor*)
GLOBAL WARMING AND CLIMATE CHANGE
Prospects and Policies in Asia and Europe

Catherine Mitchell, Jim Watson and Jessica Whiting (*editors*)
NEW CHALLENGES IN ENERGY SECURITY
The UK in a Multipolar World

Catherine Mitchell
The Political Economy of Sustainable Energy

Ivan Scrase and Gordon MacKerron (*editors*)
ENERGY FOR THE FUTURE
A New Agenda

Gill Seyfang
Sustainable Consumption, Community Action and the New Economics
Seeds of Change

Benjamin K. Sovacool
ENERGY & ETHICS
Justice and the Global Energy Challenge

Joseph Szarka
WIND POWER IN EUROPE
Politics, Business and Society

Joseph Szarka, Richard Cowell, Geraint Ellis, Peter A. Strachan and Charles Warren (editors)
LEARNING FROM WIND POWER
Governance, Societal and Policy Perspectives on Sustainable Energy

David Toke
ECOLOGICAL MODERNISATION AND RENEWABLE ENERGY

Xu Yi-chong (editor)
NUCLEAR ENERGY DEVELOPMENT IN ASIA
Problems and Prospects

Xu Yi-chong
THE POLITICS OF NUCLEAR ENERGY IN CHINA

Also by Benjamin K. Sovacool

CLIMATE CHANGE AND GLOBAL ENERGY SECURITY (*with Marilyn A. Brown*)

CONTESTING THE FUTURE OF NUCLEAR POWER

ENERGY ACCESS, POVERTY, AND DEVELOPMENT (*with Ira Martina Drupady*)

ENERGY AND AMERICAN SOCIETY (*edited with Marilyn A. Brown*)

ENERGY POVERTY

ENERGY SECURITY, INEQUALITY AND JUSTICE (*with Roman Sidortsov and Benjamin R. Jones, forthcoming*)

GLOBAL ENERGY JUSTICE (*forthcoming*)

POWERING THE GREEN ECONOMY: The Feed-in Tariff Handbook (*with Miguel Mendonca and David Jacobs*)

THE ROUTLEDGE HANDBOOK OF ENERGY SECURITY (*edited*)

THE DIRTY ENERGY DILEMMA: What's Blocking Clean Power in the United States

THE GOVERNANCE OF ENERGY MEGAPROJECTS: Politics, Hubris and Energy Security (*with Christopher J. Cooper, forthcoming*)

THE NATIONAL POLITICS OF NUCLEAR POWER: Economics, Security, and Governance (*with Scott Victor Valentine*)

Energy, Climate and the Environment
Series Standing Order ISBN 978–0–230–00800–7 (hb) 978–0–230–22150–5 (pb)
You can receive future titles in this series as they are published by placing a standing order. Please contact your bookseller or, in case of difficulty, write to us at the address below with your name and address, the title of the series and one of the ISBNs quoted above.

Customer Services Department, Macmillan Distribution Ltd, Houndmills, Basingstoke, Hampshire RG21 6XS, England

Energy & Ethics

Justice and the Global Energy Challenge

Benjamin K. Sovacool

Associate Professor of Law and Senior Researcher, Vermont Law School, USA

Professor of Social Sciences, Aarhus University, Denmark

First published 2013 by
PALGRAVE MACMILLAN

Palgrave Macmillan in the UK is an imprint of Macmillan Publishers Limited, registered in England, company number 785998, of Houndmills, Basingstoke, Hampshire RG21 6XS.

Palgrave Macmillan in the US is a division of St Martin's Press LLC, 175 Fifth Avenue, New York, NY 10010.

Palgrave Macmillan is the global academic imprint of the above companies and has companies and representatives throughout the world.

Palgrave® and Macmillan® are registered trademarks in the United States, the United Kingdom, Europe and other countries

ISBN: 978-1-137-29864-5 hardback
ISBN: 978-1-137-29865-2 paperback

This book is printed on paper suitable for recycling and made from fully managed and sustained forest sources. Logging, pulping and manufacturing processes are expected to conform to the environmental regulations of the country of origin.

A catalogue record for this book is available from the British Library.

A catalog record for this book is available from the Library of Congress.

To Lilei, who inspired me to take a closer look at the people affected by energy systems, rather than just the technology

Contents

List of Tables

List of Figures

Series Editor Preface

Concerns about the potential environmental, social and economic impacts of climate change have led to a major international debate over what could and should be done to reduce emissions of greenhouse gases. There is still a scientific debate over the likely *scale* of climate change, and the complex interactions between human activities and climate systems, but, global average temperatures have risen and the cause is almost certainly the observed build up of atmospheric greenhouse gases.

Whatever we now do, there will have to be a lot of social and economic adaptation to climate change – preparing for increased flooding and other climate related problems. However, the more fundamental response is to try to reduce or avoid the human activities that are causing climate change. That means, primarily, trying to reduce or eliminate emission of greenhouse gasses from the combustion of fossil fuels. Given that around 80% of the energy used in the world at present comes from these sources, this will be a major technological, economic and political undertaking. It will involve reducing demand for energy (via lifestyle choice changes – and policies enabling such choices to be made), producing and using whatever energy we still need more efficiently (getting more from less), and supplying the reduced amount of energy from non-fossil sources (basically switching over to renewables and/or nuclear power).

Each of these options opens up a range of social, economic and environmental issues. Industrial society and modern consumer cultures have been based on the ever-expanding use of fossil fuels, so the changes required will inevitably be challenging. Perhaps equally inevitable are disagreements and conflicts over the merits and demerits of the various options and in relation to strategies and policies for pursuing them. These conflicts and associated debates sometimes concern technical issues, but there are usually also underlying political and ideological commitments and agendas which shape, or at least colour, the ostensibly technical debates. In particular, at times, technical assertions can be used to buttress specific policy frameworks in ways which subsequently prove to be flawed.

The aim of this series is to provide texts which lay out the technical, environmental and political issues relating to the various proposed policies for responding to climate change. The focus is not primarily on the science of climate change, or on the technological detail, although there will be accounts of the state of the art, to aid assessment of the viability of the various options. However, the main focus is the policy conflicts over which strategy to pursue. The series adopts a critical approach and attempts to identify flaws in emerging policies, propositions and assertions. In particular, it seeks to illuminate counter-intuitive assessments, conclusions and new perspectives. The aim is not simply to map the debates, but to explore their structure, their underlying assumptions and their limitations. Texts are incisive and authoritative sources of critical analysis and commentary, indicating clearly the divergent views that have emerged and also identifying the shortcomings of these views. However the books do not simply provide an overview, they also offer policy prescriptions.

Acknowledgments

The Energy Security and Justice Program at Vermont Law School's Institute for Energy and the Environment investigates how to provide ethical access to energy services and minimize the injustice of current patterns of energy production and use. It explores how to equitably provide available, affordable, reliable, efficient, environmentally benign, proactively governed, and socially acceptable energy services to households and consumers. One track of the program focuses on lack of access to electricity and reliance on traditional biomass fuels for cooking in the developing world. Another track analyzes the moral implications of existing energy policies and proposals, with an emphasis on the production and distribution of negative energy externalities and the impacts of energy use on the environment and social welfare.

This book is one of three produced by the Program. The first, *Energy Security, Inequality, and Justice*, maps a series of prominent global inequalities and injustices associated with modern energy use. The second, *Energy and Ethics: Justice and the Global Energy Challenge*, presents a preliminary energy justice conceptual framework and examines eight case studies in which countries and communities have overcome energy injustices. The third, *Global Energy Justice: Principles, Problems, and Practices*, matches eight philosophical justice ideas with eight energy problems, and examines how these ideals can be applied in contemporary decision-making.

Given the novelty of this particular book, and the fact that in many instances I was learning about case studies at second hand, I am most indebted to a number of colleagues for their specific input. Professor David Elliott from the Open University graciously read and commented on a first draft of the entire book. Professor Henrik Lund from the Department of Development and Planning at Aalborg University and Professor Michael Evan Goodsite from the School of Business and Social Sciences at Aarhus University courteously provided comments on "chapter 2: Availability and Danish Energy Policy." Stephen Young from Vermont Law School and the University of East Anglia, Professor Gordon Walker from the Lancaster Environment Centre at Lancaster University, Senior Research Fellow Jan Gilbertson from the Centre for Regional Economic and Social Research at Sheffield Hallam University, and Professor Brenda Boardman from Oxford University all kindly reviewed "chapter 3: Affordability and Fuel Poverty in England." An anonymous former member of the World Bank's Inspection Panel charitably reviewed "chapter 4: Due Process and the World Bank's Inspection Panel." Professor Ann Florini from both the Brookings Institution and

Singapore Management University and Anders Tunold Kråkenes from the Extractive Industries Transparency Initiative International Secretariat kindly reviewed "chapter 5: Information and the Extractive Industries Transparency Initiative." Jan Hartman from the Earth Institute at Columbia University helpfully reviewed "chapter 6: Prudence and São Tomé e Príncipe's Oil Revenue Management Law." Deputy Chief Executive Officer S. M. Formanul Islam from the Infrastructure Development Company Limited in Bangladesh considerately reviewed "chapter 7: Intergenerational Equity and Solar Energy in Bangladesh." Dr Robert K. Dixon, Head of the Climate Change and Chemical Teams at the Global Environment Facility, and Roland Sundstrom from the Adaptation Cluster at the Global Environment Facility, generously reviewed "chapter 8: Intragenerational Equity and Climate Change Adaptation." Lavinia Warnars from the School of Social and Political Sciences of the Environment at Radboud University, and owner and founder of LES: Lavinia's Eco Solutions, thoughtfully reviewed "chapter 9: Responsibility and Ecuador's Yasuní-ITT Initiative."

Lastly, I am most grateful to Professor Aleh Cherp and the Central European University in Budapest, Hungary, for an Erasmus Mundus Visiting Fellowship with the Erasmus Mundus Master's Program in Environmental Sciences, Policy, and Management (MESPOM), which has supported elements of the work reported here.

Any opinions, findings, and conclusions or recommendations expressed in this book are my own, however, and do not necessarily reflect the views of Vermont Law School, the Institute for Energy and Environment, Central European University, or any of my colleagues who reviewed chapters and earlier drafts.

About the Author

Benjamin K. Sovacool is Visiting Associate Professor at Vermont Law School, where he manages the Energy Security and Justice Program at their Institute for Energy & the Environment. Professor Sovacool works as a researcher and consultant on issues pertaining to renewable electricity generators and distributed generation, the politics of large-scale energy infrastructure, designing public policy to improve energy security and access to electricity, and building adaptive capacity to the consequences of climate change. He is the author, editor, co-author, or co-editor of 13 books on energy security and climate change issues in addition to hundreds of peer-reviewed academic studies.

List of Acronyms and Abbreviations

ADAMS	Association for Development Activity of Manifold Social Work of Bangladesh
ADB	Asian Development Bank
AFAUS	Al-Falah Aam Unnayan Sangstha of Bangladesh
ANP	National Petroleum Agency of São Tomé e Príncipe
AVA	Association for Village Advancement of Bangladesh
BGEF	Bright Green Energy Foundation of Bangladesh
CAF	Andean Development Corporation
CAN	Andean Community of Nations
CHP	combined heat and power
CNPC	Chinese National Petroleum Corporation
COP	Conference of the Parties
DFID	Department for International Development of the United Kingdom
DG	distributed generation
DGM	Department of Geology and Mines of Bhutan
DOE	Department of Energy of the United States
EITI	Extractive Industries Transparency Initiative
EPC	Energy Performance Certificate
ELN	National Liberation Army of Columbia
EOO	European Ombudsman Office
EPA	Environmental Protection Agency of the United States
FAO	Food and Agricultural Organization
FARC	Revolutionary Armed Forces of Columbia
FIT	feed-in tariff
FPIC	free prior informed consent
GCF	Green Climate Fund
GEF	Global Environment Facility
GIZ	Deutsche Gesellschaft für Internationale Zusammenarbeit
GHEL	Green Housing and Energy Limited of Bangladesh
GJ	gigajoules
GLOF	Glacial Lake Outburst Flood
GPOBA	Global Partnership on Output-Based Aid
IADB	Inter-American Development Bank
IAEA	International Atomic Energy Agency
IAP	indoor air pollution
IBRD	International Bank for Reconstruction and Development
ICSID	International Centre for the Settlement of Investment Disputes

IDA	International Development Association
IDB	Islamic Development Bank
IDCOL	Infrastructure Development Company Limited of Bangladesh
IEA	International Energy Agency
IFC	International Finance Corporation
IMF	International Monetary Fund
IP	World Bank Inspection Panel
IPCC	Intergovernmental Panel on Climate Change
ITT	Ishpingo Tambococha Tiputini oilfields
IUCN	International Union for Conservation of Nature and Natural Resources
JDZ	Nigerian-STP Joint Development Zone
KGOE	kilograms of oil equivalent
kV	kilovolt
kW	kilowatt
kWh	kilowatt-hour
LDCF	Least Developed Countries Fund
MIGA	Multilateral Investment Guarantee Agency
MPTF	Multi-Partner Trust Fund
MW	megawatt
MWp	megawatt-peak
NAPA	National Adaptation Programs of Action
NEITI	Nigerian Extractive Industries Transparency Initiative
NPTC	National Thermal Power Corporation of India
NRFs	Natural Resource Funds
NUSRA	Network for Universal Services and Rural Advancement of Bangladesh
OECD	Organization of Economic Cooperation and Development
OPEC	Organization of the Petroleum Exporting Countries
ORML	Oil Revenue Management Law
PA	Participation Agreement
PM	particulate matter
PO	Participating Organization
PV	photovoltaic
PWYP	Publish What You Pay campaign
REB	Rural Electrification Board of Bangladesh
REDD	Reducing Emissions from Deforestation and Forest Degradation
REDI	Rural Energy and Development Initiative of Bangladesh
RDF	Resource Development Foundation of Bangladesh
SCORE	Sarawak Corridor of Renewable Energy
SEF	SolarEn Foundation of Bangladesh
SFDW	Shakti Foundation for Disadvantaged Women of Bangladesh
SHS	Solar Home System
SS	*Soura Shakti*

STP	São Tomé e Príncipe
SUV	sport utility vehicle
T&D	transmission and distribution
TJ	terrajoules
UNDP	United Nations Development Program
UNFCCC	United Nations Framework Convention on Climate Change
UNSAR	South American Union of Nations
USGS	United States Geological Survey
WF	Warm Front Scheme
WRI	World Resources Institute
YGC	Yasuní Guarantee Certificate
YNF	Yasuní National Forest

1
Introduction

Introduction

The forester and philosopher Aldo Leopold once wrote that promoting sustainable development was "a job not of building roads into lovely country, but of building receptivity into the still unlovely human mind."[1] This book argues that it's time to build some receptivity into our minds concerning the immorality and injustice of the decisions we each make about energy production and consumption.

Though we may not see it directly, [our modern reliance on oil, coal, natural gas, and uranium – the world's four main sources of modern energy – continues to involve, and at times worsens, gross inequality and inequity, and it presents pressing moral and ethical concerns.] One-quarter of humanity lives in homes without reliable and affordable access to heating, lighting, and cooking, and [there are more humans living without electricity today than at any other time in human history.] Policymakers in Europe and the United States ignore human rights abuses and the treatment of women in major oil-exporting countries to ensure security of supply and low market prices. Farmers and landowners in the rainforests of Asia and Africa degrade their land to provide fuel for cooking and support community incomes, raising standards of living but collapsing habitats and contributing to global greenhouse gas emissions. City and county officials locate trash incinerators, coal mines, and landfills near state borders when possible so that some of their pollution is transferred to other communities, a problem known as "state line syndrome." The funds from oil extraction and production in Azerbaijan and Nigeria have generated more revenue for presidential palaces, private airplanes, and prostitutes rather than bread, books, and better social services.

Energy, like many other processes and technological systems in society, can be used for good or evil. To those who hope that renewable forms of energy such as solar panels and wind turbines will by themselves emancipate us from a world of oil dependence and climate change, consider that

1

imperialism and colonialism arose during an era of renewable energy and associated technologies, which formed the energy basis of British, French, Spanish, Dutch, and Portuguese colonial empires. Wind-powered sailboats conquered the world, windmills ground sugarcane on slave plantations, and water-powered pumps drained lakes while water-powered mills accelerated deforestation.[2] Without direction concerning what is "moral" and "right," the utilization of renewable energy fuels and services can dominate and enslave as much as they can democratize and liberate.

These examples raise three key moral questions: (1) Are we being fair to present generations in giving some people disproportionate access to the benefits of energy while giving others its burdens? (2) Are we being fair to future generations in leaving a legacy of nuclear waste, the depletion of fossil fuels, and the pollution of the atmosphere and climate? (3) What do justice and ethics have to contribute to how we all make decisions about energy? This book answers these questions, as well as a few others. It connects concepts from justice theory and ethics to pressing concerns about energy policy, technology, and security. In doing so, it combines the most up-to-date data on global energy security and climate change with appraisals based on centuries of thought about the meaning of justice in social decisions.

Novelty and contribution

In traveling down this ethical road, this book differs from others on energy, the environment, and climate change in four meaningful ways.

First, and most obviously, the book critiques the view that energy policy and security problems are matters best left to economists and engineers. The book resonates strongly with the statement from philosopher and humanist Paul Goodman, who once wrote that "whether or not it draws on new scientific research, technology is a branch of moral philosophy, not of science."[3] The book therefore rejects the idea that the market will "solve" our energy problems and that scientists and engineers can "design" technical solutions without first addressing fundamental moral questions about justice and ethics. Left to their own devices, global energy markets will prolong the use of fossil fuels as long as they are profitable to extract and use, down to the last remaining drops of oil and lumps of coal, even if their combustion and use permanently damages the climate and ruins local communities, or if their benefits seemingly outweigh their costs (even if all of the benefits accrue to one wealthy company and the costs afflict thousands of penniless villagers). Similarly, research scientists and engineers will help them do so as long as they have a vested interest in the energy sector – which hundreds of thousands do.

Influenced by this flawed paradigm obsessing over economics and technology, energy analysts frequently ask the wrong questions. They will ponder how large proven reserves of oil and gas are rather than questioning the need to utilize oil and gas in the first place, or to ask whether these infrastructures are fair to their workers and the communities that live near them. They will assess and model energy prices and technological learning curves rather than ask how existing energy infrastructures benefit some people to the exclusion of others.

On the contrary, this book argues that it is a mistake to talk about building infrastructure, improving energy security, developing energy resources, forecasting future energy demand, or conducting research on new technologies without first asking what this energy is for, what values and moral frameworks ought to guide us, and who benefits. Too often, national and international energy policies have focused on protecting adequate supplies of conventional fuels with little to no regard for the long-term consequences to the people and cultures the policies are intended to benefit. For, as eminent historian David F. Noble writes, without morality "the technological pursuit of salvation can become a threat to our survival."[4]

Second, because of its focus on justice, the book relies on concepts and sources of data from philosophy, law, and ethics, along with politics, economics, sociology, psychology, and history. In doing so, the book ignores the usual dichotomies in energy scholarship and analysis such as "supply" versus "demand," "scientific" methods and disciplines versus "social science" ones, and "technology" versus "behavior."

Third, the book utilizes what geographers and political scientists call a "polycentric" lens that classifies "decision-makers" not only as policymakers and regulators but also as ordinary students, jurists, homeowners, businesspersons, and consumers. The book therefore discusses "top-down" justice solutions such as national legislation and policy mechanisms alongside "bottom-up" elements such as household energy programs and community-based climate change adaptation projects.

Fourth, the book has a comparative, global focus with case studies involving dozens of countries. This is because energy issues are now global in scope. Worldwide trade in oil and gas amounts to roughly $2.2 trillion per year and two-thirds of all oil and gas is traded internationally,[5] in addition to another $1.3 trillion in annual revenues from the extractive industries sector, to which coal is the largest contributor. Indeed, 37 percent of global carbon dioxide emissions are from fossil fuels traded internationally – that is, they were not consumed in their countries of origin – and an additional 23 percent of global emissions are embodied in traded goods and spread across a supply chain involving at least two or more countries.[6] The energy industry, its resources and shortages, its people, its prices, and its emissions are interdependent, and this interconnectedness requires that an assessment of energy justice takes a concomitant global focus.

The book is not, however, merely about abstract justice and ethics concepts. It is also about people, and, to some degree, hope. In the pages to come, you will read about how self-organized groups of cooperatives played a strong role in convincing Denmark to switch from centralized power plants to decentralized wind farms, and completely rejected nuclear power in the process. You will read about how thousands of elderly and poor British citizens literally freeze to death in the winter because they cannot afford heat, but also about how a national program has lifted millions of these homes out of fuel poverty. You will read about how an independent yet brave three-person Inspection Panel has stood up to the executive management of the World Bank and persuaded them to abandon or dramatically improve the Bank's plans for hydroelectric dams, coal mines, road transport systems, and other types of infrastructure. You will read about how countries such as Chad and Nigeria have voluntarily joined a global transparency initiative that explicitly gives civil society groups and communities a platform to raise issues over petroleum revenues. You will read how the small island country of São Tomé e Príncipe has decided to invest all of its revenue from the energy industry into a fund for social development projects and future generations. You will read how a national infrastructure bank has pumped millions of dollars into the dissemination of solar home systems in Bangladesh to the point where more than five million individuals, mostly women and children, have received affordable yet high-quality light for the first time. You will read about how a collection of rich countries such as Germany and the United States have banded together to donate half a billion dollars to least developed countries such as Bhutan and Cambodia so that they can prevent glacial floods and improve their food security in the face of climate change. You will, lastly, read about an ambitious proposal by Ecuador to keep almost one billion barrels of oil in the ground under the Yasuní National Forest so that its precious diversity, and the cultural heritage of the Huaorani, Quechua, Tagaeri, and Taromenane people, is preserved.

But, for those who want to read about more than people and hope, an exploration of energy justice is also about good business, and at making money while doing right. Three of the book's case studies are examples of how promoting energy justice brings exorbitant social and economic benefits that far outweigh its cost. In the United Kingdom, Chapter Two reveals that £2.4 billion invested to fight energy poverty has yielded £87.2 billion in savings. Chapter Eight demonstrates that the climate change adaptation measures being promoted by the Global Environment Facility have a future return on investment of 40 to 1, cut carbon dioxide emissions 2 to 500 times more cheaply than alternatives, and leverage an additional $4.66 for every $1 funneled into their Least Developed Countries Fund. Chapter Nine showcases how Ecuador's Yasuní-ITT initiative keeps $7.2 billion worth of oil undeveloped so that $32.8 billion in avoided emissions, preserved forests, and national poverty targets can be achieved.[7]

Furthermore, the book matters not only for purely philosophical reasons, but because justice can directly impact community livelihoods and the bottom line of energy corporations. In the United Kingdom, for example, a saboteur breached the most heavily guarded power station in the country, ruined one of the plant's 500 MW turbines and left a homemade poster protesting against coal. The incident forced the coal- and oil-fired facility to suspend electricity generation for four hours and caused greenhouse gas emissions over the entire UK to temporarily drop by two percent.[8] In Australia, dozens of protesters scaled 50-meter walls and chained themselves to the Hay Point Export Terminal near Mackay, forcing its closure for two days, preventing the export of 180,000 tons of coal, and causing $14 million in lost revenues. In Nepal, Maoist rebels repeatedly bombed and attacked various hydroelectric power stations in an effort to force the government to promote political pluralism and reduce corruption.[10] In the Sao Paulo state of Brazil, 900 women from Via Campesina occupied the Cevasa sugar mill to protest against its labor practices, suspending operations for a week and inducing substantial economic losses.[11] In the Port of Hamburg, Germany, the *Klimacamp* group sent 200 people to blockade the world's largest refinery for biodiesel operated by Archer Daniels Midland, shutting down the refining of Indonesian palm oil.[12] In the Philippines, Communist New People's Army rebels raided a state-owned plantation used for the manufacturing of biofuels from jatropha on Negros Island, where they torched equipment and stopped workers from hauling lumber.[13] In the Guangdong province of China, police shot and killed as many as 20 people for protesting against lack of compensation for land for local wind farms.[14] Single conflicts such as these can cost companies millions of dollars in delays, lawsuits, missed opportunities, social dislocation, and the damage of corporation reputations.[15]

Some of these actions may be seen as alarming, and some were certainly illegal. But the participants clearly felt they were necessary. The point is that, when people feel that energy justice has been violated, they can disrupt operations, boycott brands, picket companies, and even attack and destroy energy infrastructure and equipment. But, equally, sometimes doing what is just makes economic sense. Either way, energy justice is not merely an academic concern.

The global energy system

Before delving into how ethics and justice can guide energy decisions, it's useful to first provide some basic statistics about the global energy system, which is capital-intensive, expensive, and expansive. From 1900 to 2000 the population of the earth quadrupled from 1.6 billion to 6.1 billion, but annual average supply of energy per capita grew *even more*, from 14 gigajoules (GJ) in 1900 to roughly 60 GJ in 2000. Over this period, energy consumption in the US more than tripled, quadrupled in Japan, and increased by a factor of 13 in China. Individual examples of more energy-intensive lifestyles are striking. In 1900, an affluent farmer in the Midwest of the US holding the reins of six large horses plowing a field would generate about 5 kW of animate power.

A century later, that same type of farmer could sit in a large tractor with an air-conditioned and stereo enclosed cabin and harness 250 kW. In 1900, an engineer operating a locomotive could reach speeds close to 100 km/hour with about 1 MW of steam power, yet in 2000 the pilot of a Boeing 747–400, 11 km above the Earth's surface, could exceed 900 km/hour and a discharge of 120 MW.[16]

These modern patterns of energy use reflect a fundamental transition from principal sources of energy derived directly from the sun (such as human and animal muscle power, wood, flowing water, and wind) to those dependent on fossil fuels. Global use of hydrocarbons as a fuel by humans, for example, increased 800-fold from 1750 to 2000 and 12-fold again from 1900 to 2000.[17] Given that the average human body produces between 60 and 90 W of equivalent energy per hour, it would take 13.3 billion people – almost twice as many as exist on Earth – to naturally produce as much energy as America's electricity grid does in 60 minutes.[18] Some estimates have calculated that the global workforce of labor is roughly three billion workers, and that 21 percent of that workforce is engaged in industrial activities directly connected to energy extraction, production, and consumption – and the resulting number of 630 million workers excludes those employed in energy-intensive sectors such as agriculture or the construction of buildings.[19]

Consequently, modern energy production and use involve multiple fuel chains that cut across vertical scales within a country. A prime case would be a coal system that involves the coal mine and railway as well as the power plant and transmission and distribution network; or a wind farm which requires the production of aluminum, copper, concrete, and fiberglass "upstream" to make the turbines and other components as well as switching stations and interconnection to the electricity network "downstream" from the turbines themselves. The extractive industries and mining sector overlaps with parts of the electricity sector, providing raw fuels such as crude oil, unprocessed natural gas, and unwashed coal through a series of hundreds of thousands of mines, onshore wells, and offshore drilling platforms, to say nothing of the material needs – copper, rare earth elements, alumina, and others – needed to manufacture power plants, transmission lines, and other electronic devices.[20] As Figure 1.1 shows, humanity used about 500 exajoules (EJ) of energy in 2005, but demand for energy will grow 45 percent between now and 2030, and this amount will almost triple by the end of the century. In 2011, the world also received a sobering 81 percent of its energy from fossil fuels.[21] Though the specific technological and fuel sources vary, the global energy system primarily consists of three interconnected yet distinct sectors: electricity, transport, and heating and cooking.

Electricity

Reliance on electricity has grown significantly due to instant and effortless access, its easily adjustable flow, facilitation of high-precision speed and

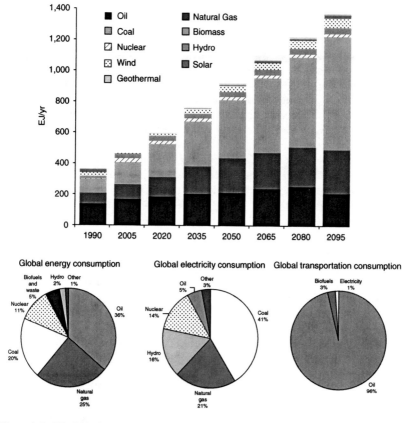

Figure 1.1 Worldwide energy use, 1990–2095
Source: International Energy Agency and US Energy Information Administration.

process controls, cleanliness, and silence at the point of use. Electricity now powers not only lights, refrigerators, televisions, and radios but also vehicles, electric fireplaces, electric arc furnaces, and movable sidewalks. In 1900 less than two percent of the world's electricity came from fossil fuels, but by 1950 the number had jumped to 10 percent, and passed 67 percent in 2010. Humans have also become more dependent on information and media technology such as the internet and televisions, which have corresponded with a dramatic increase in the manufacturing of information storage, telecommunications, and electronics as well as the energy needed to operate such devices.[22]

Thus, as of early 2012, roughly 170,000 generators provided electricity at more than 75,000 power plants – about half of them coal-fired, 440 of them nuclear-powered – and they transmitted electricity through roughly four million miles of transmission and distribution lines. The longest transmission line in the world, in the Democratic Republic of the Congo, spans more than 1,056 miles and cost $900 million to build. In the United States the electricity sector is so big that it consists of almost 20,000 power plants, half a million miles of high-voltage transmission lines, 1,300 coal mines, 410 underground natural gas storage fields, and 125 nuclear waste storage facilities, in addition to hundreds of millions of transformers, distribution points, electric motors, and electric appliances. It is the most capital-intensive sector of economic activity for the country and represents about 10 percent of sunk investment. Expenditure on electricity reached $355 billion in 2007 (3.2 percent of the nation's GDP in that year),[23] and the country had more electric utilities and power providers than *Burger King* restaurants.[24] Particular forms of electricity supply, such as nuclear power, differ further still by necessitating their own collection of facilities unique to the fuel cycle, such as uranium mines, uranium mills, enrichment facilities, fuel cladding facilities, temporary waste sites, permanent waste repositories, research laboratories, and research reactors.

Electricity has transformed industry and society by improving the productivity and quality of life of populations and regions around the world. Nearly every aspect of daily life in a modern economy depends on electricity. Electricity's extraordinary versatility as a source of energy means it can be put to an almost limitless use to heat, light, transport, communicate, compute, and generally power the world. The backbone of modern industrial society is, and for the foreseeable future can be expected to remain, the use of electricity.

Transport

The introduction of mobile engines and inexpensive liquid fuels has enhanced personal mobility and facilitated new modes of transport. The number of mass-produced motorized vehicles jumped from a few thousand in 1900 to more than 700 million in 2000 (and about one billion today) along with a notable increase in vacation travel and nonessential trips. (In 1900, for instance, there were only 8,000 cars in the United States and just 15 kilometers of paved roads.[25]) The mobility of people has been matched by the increasing movement of goods and services as trade and commerce have accelerated. In 2000, trade alone accounted for 15 percent of global economic activity, most of it through diesel-powered trains, trucks, and tankers.

The rise of the coveted automobile is often characterized as one of the great achievements of the twentieth century. During the first half of the century, the gasoline-powered vehicle evolved from a fragile, cantankerous, and faulty contraption to a streamlined, reliable, fast, luxurious, and widely affordable product.[26] These automotive engineering feats were enhanced by the creation of interstate highway systems and urban infrastructure that have offered many people unprecedented mobility.

The transport system that resulted must deliver more than 20 million barrels of oil per day to the United States alone, or 55 percent of the world's 87.8 million barrels of oil per day globally (84.8 million barrels per day in 2006, as Table 1.1 shows), backed by more than one thousand refineries and almost one million gasoline stations, to the world's roughly one *billion* automobiles which drive on 11.1 million miles of paved roads – enough to drive to the moon and back 46 times.[27] These roads require $200 million of maintenance per day, and constitute a paved area equal to all arable land in Ohio, Indiana, and Pennsylvania. Indeed, Figure 1.2 shows that world oil consumption has almost tripled over the past five decades, growing from about 30 million barrels per day in 1960 to almost 90 million barrels per day in 2010. The transport system also creates seven billion pounds of un-recycled scrap and waste each year. The United States has a voracious appetite for oil, with consumption having climbed 25 percent since the mid-1980s. China is also learning to drink deep, and may be using an equivalent amount to the US in a relatively short timeframe.

Consumers worldwide have precious little choice surrounding their fuels for transport: human civilization relies on crude oil to meet 96 percent of its transportation demand (electricity, biofuels, and a small amount of natural gas meet the rest).[28] In the developed world, the car maintains its enduring stranglehold over public transportation options. Statistics agencies in many North American states and provinces report that over 80 percent of their workforce commutes by car. In other words, gasoline vehicles have become a transportation monoculture – cars and petroleum dominate. Automobile

Table 1.1 Global oil demand, 2006 and 2025

Country	Barrels/Day (Bpd) in 2006	Per Capita/ Day (Gallons)	Projected 2025 (Bpd)	GHG Emissions 2003 (Tg CO2e) from Transport
United States	20.92 Million	2.87	28.3 Million	6,072.18
China	7.03 Million	0.17	12.8 Million	4,057.32
Japan	5.34 Million	1.76	5.8 Million	1,339.19
Canada	2.24 Million	2.67	2.8 Million	696.26
Europe	15.45 Million	1.38	17 Million	3,872.96
Brazil	1.8 Million	0.44	3.8 Million	658.97
Latin America	5.21 Million	0.74	6.4 Million	383.07
Middle East	6.47 Million	0.34	6.9 Million	649.57
Africa	2.95 Million	0.04	3.1 Million	362.17
India	2.24 Million	0.08	5.3 Million	385.43
World	84.80 Million	–	96 Million	34,580.34

Tg CO2e, billion tons of carbon dioxide equivalent
Source: Adapted from BP Statistical Review of Energy.

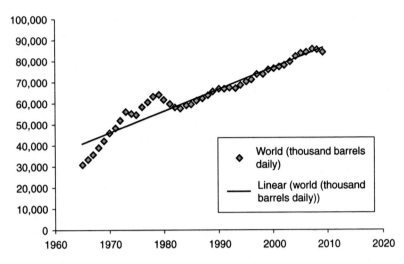

Figure 1.2   World oil consumption (thousand barrels per day), 1960–2010
Source: Adapted from BP Statistical Review of Energy.

domination is evidenced by the fact that public transport accounts for only about three percent of passenger travel in the US,[29] and rail transport accounts for less than one percent.[30] Gasoline domination is evidenced by the fact that today nonpetroleum fuels (including electricity, biofuels, and natural gas) account for only four percent of US transportation fuel consumption (up from two percent in 2007).[31] The motorization of America has resulted in more than one automobile for every licensed US driver.

In the developing world, dirt-cheap cars are gaining momentum in a marketplace of billions. Throughout India, China, and Indonesia middle-class families are laden three or four to a single scooter, choking through diesel exhaust on overcrowded roads. Although 85 percent of the world's population do not have access to a car, they aspire to car ownership, especially residents of the rapidly growing South and East Asian nations. In China, the conventional vehicle fleet is expected to grow tenfold, from 37 million in 2005 to 370 million by 2030.[32] In total, within a couple of decades, the world is projected to have two billion gasoline-powered automobiles – twice the number that currently exist.[33] If China continues its car-centric development model, it could by itself add another billion cars by the end of the century.

Heating and cooking

Though electricity accounts for roughly 17 percent of global final energy demand, low-temperature heat accounts for about 44 percent. Globally, this means that people use more energy for heating, cooking, and thermal comfort – typically by burning woody biomass – than for any other purpose.[34]

For instance, the majority of households in the developing world – about three out of every four – rely on traditional stoves for their cooking and heating needs. Traditional stoves range from three-stone open fires to substantial brick and mortar models and ones with chimneys.[35] Because these stoves are highly inefficient (as much as 90 percent of their energy content is wasted), they require a significant amount of fuel – almost two tons of biomass per family per year. These consumption patterns strain local timber resources and can cause "wood fuel crises" when wood is harvested faster than it is grown.[36] The burning of this biomass also exacerbates climate change. Households in developing countries burn about 730 tons of biomass annually, which translates into more than one billion tons of carbon dioxide.[37] As an illustration, Japan, currently the fifth largest carbon dioxide emitter in the world, emitted 1.2 billion tons of carbon dioxide in 2008.[38]

Billions of people consequently rely on wood, charcoal, and other biomass fuels to meet their daily energy needs. Such fuels comprise 40 to 60 percent of total energy consumption for many communities, and household cooking is responsible for 60 percent of total energy use in Sub-Saharan Africa (and exceeds 80 percent in some countries). Poor families spend one-fifth or more of their income on wood and charcoal, devote one-quarter of household labor to collecting fuelwood, and then suffer the life-endangering pollution that results from inefficient combustion.[39]

Energy justice and the chapters to come

Clearly, these statistics imply that the global energy system is on the wrong track, and various independent assessments have confirmed the overall deterioration of different energy metrics and indicators. For instance, one scorecard of energy sustainability performance in the United States found that the country has backslid from the 1970s to today. Only in two out of ten metrics – energy intensity, and sulfur dioxide emissions – did the country improve.[40] A similar survey of 22 countries in the Organization of Economic Cooperation and Development (OECD) found equivalent results: Denmark, with the best score, led the pack with a score of 3 out of a highest possible score of 10. A majority of countries did poorly, with 13 countries scoring below zero, implying that their energy security has worsened from 1970 to 2007.[41] A third assessment of countries in the Asia Pacific utilized 20 distinct indicators spread broadly across the areas of energy supply, energy affordability, efficiency and innovation, environmental stewardship, and governance.[42] The best possible score a country could have gotten – if it excelled in every category, for every year – was 500, yet the top performer (Japan) scored only 284.[43]

Thankfully, the concept of energy justice provides a framework for how to minimize and at times eliminate the various energy threats and injustices confronting countries and communities. Spelled out in the next eight chapters, and synthesized in Chapter Ten, this framework is centered on eight key principles:

- Availability, which holds that people deserve sufficient energy resources of high quality;
- Affordability, which holds that all people, including the poor, should pay no more than 10 percent of their income for energy services;
- Due Process, which holds that countries should respect the rule of law and human rights in their production and use of energy;
- Information, which holds that all people should have access to high-quality data about energy and the environment and fair, transparent, and accountable forms of energy decision-making;
- Prudence, which holds that energy resources should not be depleted too quickly;
- Intergenerational equity, which holds that people have a right to fairly access energy services;
- Intragenerational equity, which holds that future generations have a right to enjoy a good life undisturbed by the damage our energy systems inflict on the world today; and
- Responsibility, which holds that all nations have a duty to protect the natural environment and minimize energy-related environmental threats.

The chapters to follow tease out each of these eight justice principles and apply them to real-world energy problems, followed by an in-depth case study of how a country, community, or institution promoted energy justice.

Chapter Two on "Availability and Danish Energy Policy" illustrates how dependence on other countries for energy fuels and unreliable energy infrastructure can perpetuate energy injustice. The chapter also, however, advances three practices that can enhance the availability of energy: energy efficiency, distributed generation, and renewable sources of electricity. The chapter concludes with a case study of energy policymaking in Denmark, a country that has reduced its dependence on foreign sources of energy to zero and become self-sufficient in its own energy production and use. At the core of Denmark's successful approach is a commitment to energy efficiency, prolonged taxes on energy fuels, electricity, and carbon dioxide, and incentives for combined heat and power (CHP) and wind turbines. Through these commitments, Denmark transitioned from being almost 100 percent dependent on imported fuels such as oil and coal for their power plants in 1970 to becoming a net exporter of fuels and electricity today. The country leads the world in terms of exportation of wind energy technology, with a hold on roughly one-third of the world market for wind turbines. It was able to phase out the use of virtually all oil-fired power plants in less than five years and implemented a progressive moratorium on future coal-fired power plants in the 1990s. Its most recent strategy seeks to achieve 30 percent of total energy supply from renewable energy by 2025 and 100 percent by 2050.

Chapter Three on "Affordability and Fuel Poverty in England" documents how millions of homes around the world suffer from "fuel poverty," commonly defined as the necessity to spend more than ten percent of their income paying energy bills. The chapter discusses how home energy efficiency schemes, such as those that pay to weatherize doors and windows, install insulation, and give free energy audits, can significantly reduce the prevalence of fuel poverty. The chapter finishes with an examination of the "Warm Front" program in England, which saw 2.3 million "fuel-poor" British homes receive energy efficiency upgrades to save them money and improve their overall health. Warm Front has not only lessened the prevalence of fuel poverty; it has also significantly cut greenhouse gas emissions, produced an average extra annual income of £1,894.79 per participating household, and reported exceptional customer satisfaction, with more than 90 percent of its customers praising the scheme.

Chapter Four on "Due Process and the World Bank's Inspection Panel" demonstrates how the energy system erodes human rights and undercuts community wellbeing in numerous ways, from flawed impact assessments to the forcible relocation of communities. The chapter discusses how an independent inspection panel can hold actors accountable for the ways their energy investments damage people. The chapter finishes with an analysis of the World Bank's Inspection Panel, which has improved not only its internal governance but also led to the drastic redesign of energy projects and resulted in broader positive change. Among other successes, it convinced investors and the Bank to pull out of a potentially devastating $800 million hydro-electric dam project in Nepal, and pushed for the eventual suspension of a similarly large, and socially destructive, project in India. The Panel has provided a much-needed forum for those directly affected by World Bank projects to voice their concerns, and it has demanded that the institution face the human and environmental impacts of its decisions.

Chapter Five on "Information and the Extractive Industries Transparency Initiative" documents how poor management and high levels of corruption around the world impede access to accountable, transparent information about energy reserves. In extreme cases, officials and corporate leaders divert energy resources and revenues to groups perpetrating acts of violence against indigenous people and those opposed to fossil fuel production. The Extractive Industries Transparency Initiative (EITI) provides "an independent, internationally agreed upon, voluntary standard for creating transparency in the extractive industries."[44] The EITI forces companies and governments to disclose the revenues they receive from petroleum and mining production to public groups and members of civil society. So far, 18 countries have fully complied with EITI guidelines and a further 19 countries are working to achieve EITI validation, in addition to 60 of the largest oil, gas, and mining companies in the world. Furthermore, EITI has the support of more than 80 global investment and lending institutions that manage more than $16 trillion in energy assets. EITI has substantially benefitted governments by enabling them to follow an internationally recognized transparency standard, companies by leveling the playing field and rooting out "bad apples," and citizens and members of civil society by demanding stakeholder participation in energy planning and producing high-quality information about the oil and gas sectors.

Chapter Six on "Prudence and São Tomé e Príncipe's Oil Revenue Management Law" notes how oil, natural gas, coal, and uranium took more than two billion years to accumulate, yet face depletion at a rate that will have them completely exhausted within one to two centuries, possibly sooner. Moreover, it reveals how their production, extraction, and processing have occurred with devastating consequences for national governments and local communities. Nonetheless, the chapter argues that the Oil Revenue and Management Law from the small African island country of São Tomé e Príncipe (STP) offers an excellent case study of how to create independent institutions to oversee oil development and generate as much as $20 billion – hundreds of times the country's current Gross Domestic Product (GDP) – into a National Oil Account that stipulates spending on public projects that reduce poverty and a Permanent Reserve Fund that saves for future generations after the oil runs out. STP's policies have generated much-needed government revenue, helped diversify the economy, lowered inflation and rates of poverty, and minimized corruption and the exploitation often associated with oil production.

Chapter Seven on "Intergenerational Equity and Solar Energy in Bangladesh" demonstrates that billions of people around the world – roughly one in six – lack access to electricity and almost two out of five depend on wood, charcoal, and dung for their household energy needs. This lack of access to modern energy is grossly inequitable, and it also limits income generation, blunts efforts to escape poverty, impacts the health of women and children, and contributes to global deforestation and climate change. The chapter then shows how *Soura Shakti* (SS) offers a blueprint for how planners around the world can rapidly expand access to modern energy services through the use of solar home systems. SS is the homegrown solar home systems (SHS) program of Bangladesh's Infrastructure Development Company Limited (IDCOL). The program has successfully distributed 1.42 million SHS units throughout Bangladesh and is on target to reach four million by the end of 2015. As a result of this program, Bangladesh has nine times the residential solar capacity of the United States, and it is home to the fourth largest market for photovoltaic (PV) installations after Germany, Japan, and Spain. In 2009, more than 99 percent of all SHS installations in Bangladesh occurred as part of the SS program. Even more impressive, the program achieved its initial targets three years ahead of schedule without a single customer defaulting on a loan and at $2 million below cost.

Chapter Eight on "Intragenerational Equity and Climate Change Adaptation" depicts how humanity's current emissions of greenhouse gases are changing the climate in ways that perpetuate natural disasters, intensify disease epidemics, and aggravate water scarcity, depriving future persons of the ability to meet their basic needs and live happy and productive lives. It then shows how the implementation of climate change adaptation projects gives communities the ability to improve infrastructural, institutional, and social resilience to minimize the vagaries of a harsher climate and more severe weather. The chapter presents the case study of the Global Environment Facility's Least Developed Countries Fund (LDCF), a scheme whereby industrialized countries have disbursed $602 million in voluntary contributions, raised more than four times that amount in co-financing, and supported 88 adaptation projects across 46 least developed countries. These projects have done things like accelerate coastal afforestation efforts in Bangladesh, reduce the risk of glacial lake outburst floods in Bhutan, diversify the agricultural base and improve food security in Cambodia, and improve natural disaster planning in the Maldives.

Chapter Nine on "Responsibility and Ecuador's Yasuní-ITT Initiative" focuses intently on the negative externalities associated with oil exploration and production, including blighted landscapes, poisoned atmospheres, degraded and sickened communities downstream, and emissions of particulate matter and greenhouse gases upstream. Ecuador's radical Yasuní-ITT Initiative would leave almost one billion barrels of heavy crude oil locked in perpetuity beneath one of the most intact and diverse nature reserves on the planet in order to protect biodiversity, respect the territory of indigenous peoples, combat climate change, and encourage more sustainable economic development. The Yasuní-ITT proposal has the international community pay Ecuador $3.6 billion – roughly half the value of the oil found there – in exchange for not developing the Ishpingo Tambococha Tiputini (ITT) oilfields. Funds raised so far have been placed into social and environmental development programs, including "massive programs of reforestation and natural forest recovery"[45] and the promotion of renewable energy and energy efficiency domestically.

Chapter Ten "Conclusion: Conceptualizing Energy Justice" draws inductively from each of the eight case study chapters to present a preliminary energy justice conceptual framework. This framework argues that energy justice consists of eight interrelated factors: availability, affordability, due process, good governance, prudence, intergenerational equity, intragenerational equity, and responsibility. The chapter also stresses the need for further research on the topic of energy justice and highlights the underlying moral and ethical aspects behind energy decision-making.

2
Availability and Danish Energy Policy

Introduction

The availability of energy fuels and services to those who need them is a prerequisite for other energy justice concerns. Without a supply of energy in sufficient quantity and quality, achieving other justice goals such as equity or responsibility is impossible. After justifying availability as a central justice principle, this chapter articulates how dependence on foreign suppliers as well as blackouts and unreliable energy infrastructure threaten availability. It examines the case of Denmark, which has relied upon a progression and variety of policy mechanisms such as energy taxes, research subsidies, and feed-in tariffs to promote energy efficiency, combined heat and power, and wind energy.[1]

To list just some of the most significant achievements to date, Denmark transitioned from being almost 100 percent dependent on imported fuels such as oil and coal for its power plants in 1970 to becoming a net exporter of fuels and electricity today. The country leads the world in terms of exportation of wind energy technology, with a hold on roughly one-third of the world market for wind turbines. Denmark has the largest portfolio of wind projects integrated into its power grid of any country in the world (more than 20 percent), and some parts of the country, such as Western Denmark, frequently supply more than 40 percent of their electricity from wind turbines. Combined heat and power units provide roughly 60 percent of the country's electricity and 80 percent of its heat. These achievements are all the more impressive when readers consider that the size and population of the country are roughly the same as those of the state of Maryland in the US.

Availability as an energy justice concern

The principle of availability – perhaps the least controversial part of an energy justice framework – means that sufficient energy resources, of high quality, when needed, are a cornerstone of realizing energy justice, cutting across the energy policy concerns of security of supply, sufficiency, and reliability. Indeed, the "classic" conception of energy security addresses the relative safety and source diversification of energy fuels and services with the term known as "security of supply."

The reasons for this emphasis on supply are valid, as part of ensuring availability entails procuring an adequate and uninterrupted supply and minimizing foreign dependency on imported fuels. Dependency can be costly, as most recently illustrated by Russian efforts to negotiate natural gas prices in Europe. Russia was able to successfully triple the price of natural gas exported to Belarus and Ukraine because these countries were completely dependent on Russian exports. In more serious cases, growing dependency or perceived scarcity of domestic energy supply has precipitated international conflict, including both World Wars of the twentieth century and, according to some, both of the US invasions of Iraq.[2]

Another part of availability encompasses diversification, or promoting a diversified collection of different energy technologies and ensuring variety, balance, and disparity in the energy sources and fuels employed. Diversification insulates national energy sectors from events including natural disasters, earthquakes, volcanic eruptions, tidal waves, severe weather, hurricanes, blizzards, droughts, tornadoes, landslides, lightning, dust storms, floods, routine snowfalls, spring thaws, and heavy rain. It can also foster innovation and experimentation, hedge against uncertainty and ignorance, mitigate the effects of technology lock-in, and accommodate a disparate array of values and interests.[3]

Sometimes, the decentralization of energy provision can enhance availability and reliability. Decentralization involves reliance on small-scale sources of energy supply near the point of consumption. Smaller energy projects require less interest and have more rapid construction times than larger plants. They also can generate energy closer to end-users, minimizing transmission losses and/or enabling, for example, district heating or co-generation, which improves efficiency, and facilitating better pollution controls since the volume of emissions or waste is less. Decentralizing energy supply gives local communities more incentive to insist that facilities run cleanly or provide benefits back to the neighborhood.

Unfortunately, the sufficient availability of energy is frequently under threat from dependence on foreign suppliers (prone to unexpected interruptions, transfers of wealth, and rising costs) as well as unreliable energy infrastructure. The most elemental threat to availability is interruptions in supply. To cite a few prominent examples, oil transit supplies were disrupted in Latvia in 1998, natural gas pipelines were shut off between Russia and Ukraine in 2005 and 2006, shortages of coal occurred in China in 2007, lack of rain and snow caused hydropower shortfalls in California, Brazil, and Fiji, blackouts hit North America in 2003 and Europe in 2005, and Hurricane

Katrina caused a substantial number of refinery shutdowns.[4] A multitude of serious oil supply shocks, averaging eight months in duration and affecting almost four percent of global supply, have occurred from 1950 to 2003.[5] One study identified five major disruptions in the global oil market in the two decades *after* the famed oil shocks of the 1970s:

- the Gulf War of 1990 and 1991, which removed 4.3 million barrels per day (mbd) of oil production from the market;
- suspension of Iraqi oil exports in 2001, which removed 2.1 mbd;
- a Venezuelan strike in 2003 and 2004, which removed 2.6 mbd;
- the Gulf War of 2003, which removed 2.3 mbd;
- Hurricane Katrina in 2005, which removed 1.5 mbd.[6]

In November 2007, an accidental explosion at the Enbridge Pipeline in Minnesota spilled 15,000 gallons of oil, killed two employees, shut down one-fifth of US oil imports for days and resulted in an increase in global oil prices of $4 per barrel.[7]

In addition, dependence for energy fuels on countries like Saudi Arabia or Iran transfers wealth to them that can then be used to support terrorism, condone human rights abuses and religious intolerance, or fund extremist movements. One study from the Center for Naval Analysis calculated that Iran's energy exports of $77 billion per year in 2008 provided 40 percent of the funding behind global terrorist groups, including materials and weapons to Hezbollah, insurgents in Iraq, and their own covert nuclear program. The same study estimated that Saudi Arabian organizations enriched from the country's $300 billion in annual oil revenues were able to fund a global network of extremist movements.[8] This caused one general to famously remark that, due to its demand for oil, the United States was in essence funding "both sides" of the war on terror.

Furthermore, dependence on imported sources of energy subjects importing countries to macroeconomic shocks. Researchers from Oak Ridge National Laboratory estimated that from 1970 to 2004 American dependence on foreign supplies of oil has cost the country $5.6 to $14.6 trillion, or more than the costs of all wars fought by the country going back to the Revolutionary War, including both World Wars and the first invasion of Iraq.[9] This is not the cost of the oil itself, but the direct economic costs of macroeconomic shocks and transfers of wealth. As Figure 2.1 illustrates, the lowest end of this range – $5.6 trillion – is far more than the cost of the second US invasion of Iraq, the global annual narcotics trade, the US space shuttle program, and the annual weapons trade combined.

Such figures may strike some readers as heresy, but a follow-up estimate calculated that the costs of oil dependence in the US were in excess of $500 billion for 2008 alone.[10] Each year the United States purchases more than 60 percent of its oil from foreign sources, and the cost of imported oil and

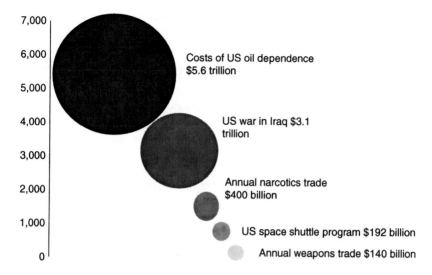

Figure 2.1 Macroeconomic costs of US dependence on foreign oil (billions of dollars), 1970–2004

refined products is the single largest contributor –48 percent – to the country's $700 billion trade deficit.[11] Another assessment recently calculated that the costs of military forces in the Persian Gulf needed to protect oil assets and infrastructure range from $50 billion to $100 billion per year;[12] a second, independent study put the figure at between $29 billion and $80 billion per year.[13] Perhaps for these reasons, energy analyst Gal Luft has remarked that "because the American way of life is one of the most energy-intensive in the world, U.S. oil dependence is a source of great national security threats."[14] As much as 40 percent of the US military budget can be attributed to protecting the oil trade.

Instead of reducing reliance on imported oil, the global economy is becoming more dependent on it. Global dependence on Middle Eastern crude oil is expected to jump from 58 percent today to 70 percent by 2015. By 2030, the International Energy Agency (IEA) predicts that the oil import dependence of India, Europe, China, and the rest of Asia will exceed 75 percent, up 10 to 25 percent over today's levels.[15] Asia's oil use already exceeds that of the US at more than 19 million barrels per day, and India and China are projected to double their consumption of oil by 2025.

The issue of oil dependence is just as stark for many nonindustrialized countries. Rises in the cost of crude oil and gasoline mean that the foreign exchange required for oil imports creates a heavy burden on their balance of trade. While developed countries spend just one or two percent of their GDP on imported oil, those in the developing world spend an average of 4.5

to nine percent of their GDP on crude oil imports. Higher prices for oil also hit developing countries twice: once for costlier barrels of oil, and again for inflated transportation costs that reflect the increase in fuel prices.[16] One study looked at the world average price of crude oil for 161 countries from 1996 to 2006, when prices increased by a multitude of seven, and concluded that lower-middle-income countries were the most vulnerable, followed by low-income countries, even though these countries consumed less oil per capita than industrialized or high-income countries.[17] The reason is that lower-income countries spent a greater share of their GDP on energy imports.

Nor are the dangers of dependence limited to oil. The electricity sectors in the United States and European Union, among others, have become dependent on foreign sources of natural gas. As suppliers in countries such as Algeria, Indonesia, Qatar, Russia, and Venezuela come to account for a greater proportion of global natural gas supply, the patterns (and subsequent costs) associated with import dependence could mimic trends in the oil sector.

A final threat to availability is declining reliability of energy infrastructure – connected in part to always trying to meet increases in demand by building more technology (rather than promoting energy efficiency), in part to the centralization and capital intensity of conventional energy systems.

Much of our existing energy infrastructure is inherently vulnerable to deliberate and accidental disruptions. Systems operators, planners, and energy firms still adhere to a classic mentality of building tightly coupled, centralized, capital-intensive forms of energy supply that can easily be disrupted by changing weather, small animals, balloons, rocks, and bullets. Electricity suppliers, for example, have clustered generating units geographically near oil fields, coal mines, sources of water, demand centers, and each other, interconnecting them sparsely and thereby making them heavily dependent on a few critical nodes and links. Operators have provided little storage to buffer successive stages of energy conversion and distribution, meaning that failures tend to be abrupt and unexpected rather than gradual and predictable. Companies and firms have located generators remotely from users, so that supply-chain links are long and the overall system lacks user controllability and comprehensibility.[18]

This complexity and interdependence can be a curse as much as a blessing, for systems can interact in unforeseen and unplanned ways. Some systems, dependent on multiple fuels or providing multiple services at once, can fail due to lack of any one of them, such as a furnace that burns oil or natural gas malfunctioning during a blackout because it needs electricity to ignite the fuel, gasoline pumps doing the same because they rely on electricity, or the outage of a microhydro plant affecting both electricity generation and mechanical processing. Others are physically dependent, meaning that the lack of one affects the other; for example, municipal water treatment systems need lots of electricity, but thermoelectric power plants need water. When built too closely together, failures in one system can cascade to the other. Broken water mains, for instance, can short out circuits or electric cables; fires and explosions can ignite entire pipeline networks; earthquakes cause gas mains to rupture and explode, destroying facilities that survived the initial shock, or cause tsunamis to wash away backup generators at nuclear reactors in Japan. In the United Kingdom, when Britain converted to natural gas from the North Sea, some public telephone booths actually started exploding. Higher gas pressure rose too quickly for older seals, causing gas to leak into adjacent telephone cable conduits, seep into booths, and be set ablaze by phone users smoking cigarettes.[19]

One sign of increased vulnerability is a greater incidence of severe blackouts and power outages. Between 1964 and 2005, for instance, the Institute of Electrical and Electronics Engineers estimates that no fewer than 17 major blackouts have affected more than 195 million residential, commercial, and industrial customers in the United States, with seven of these major blackouts occurring in the most recent ten years. Sixty-six smaller blackouts (affecting between 50,000 and 600,000 customers) occurred from 1991 to 1995 and 76 occurred from 1996 to 2000, and that's just for the United States. The costs of these blackouts are monumental: the US Department of Energy (DOE) estimates that power outages and power quality disturbances cost customers as much as $206 billion annually, or more than the entire nation's electricity bill for 1990. In the developing world, countries tend to lose one to two percent of GDP growth potential due to blackouts, overinvestment in backup electricity generators, and inefficient use of resources.[20] Nigerians live with such persistent power outages that one government official characterized the power supply as "epileptic."[21] The Nepal Electricity Authority, the state-owned monopoly supplier of power for the country, supplies electricity to the nation's capital Kathmandu for less than eight hours per day, with load shedding accounting for the remaining 16 hours.[22] In the bazaars at night, more shops rely on candles than electric bulbs to light their wares.

The simplest and best way to enhance the availability of energy services, and to minimize dependence and strengthen the resilience of energy infrastructure, is energy efficiency. Generally, the term energy efficiency refers to the reduction of energy use as the result of improved performance, increased deployment of more efficient equipment, or the alteration of consumer habits. It can include substituting resource inputs or fuels, changing preferences, or altering the mix of goods and services to demand less energy. It is a resource historically proven, through decades of experience and thousands of programs, to be the most cost-effective way of responding to increases in demand. Energy-efficiency measures have shorter lead times, need lower capital requirements, and offer consumers and utilities quick returns on investment compared with building conventional power plants, new oil and gas fields, and coal mines. Energy-efficiency technologies can improve load factors, increase cash flows, lower capital expenditure, and minimize financial risk associated with excess generating and transmission capacity.

To those who scoff at these ideas, consider that far more severe actions have been taken to promote energy efficiency at other times and in other places. After the energy crises of the 1970s, regulators in France took energy efficiency so seriously that they created the equivalent of an "energy police" to stalk the streets at night. These vanguards of conservation would issue fines to people who left their lights on or left their cars idling when they ran inside. In contemporary Beijing, China, a 20-member "energy-saving police team" monitors office buildings, schools, and hotels to make sure their heating and cooling systems do not exceed government standards (which specify that a building cannot be colder than 79 degrees Fahrenheit in the summer or warmer than 68 degrees in the winter).

Because energy efficiency cannot solve every energy problem – especially in emerging economies where energy access is still limited – it can be coupled with supply-side investments in distributed generation, combined heat and power, and renewable sources of electricity. Distributed generation (DG) refers to power generators that make electricity on-site, in small increments, close to the end-user. The technique emphasizes the deployment of small-scale generating facilities, often having installed capacities of a few to a hundred kW. DG technologies tend to be owned not just by utilities or traditional power providers but also by residential owners, commercial enterprises, and industrial firms. DG units can provide power for almost any objective and from any location. They can utilize many modes of operation, running as base-load generators, peaking units, or emergency backup facilities. They can be placed almost anywhere on the transmission system, from the low-voltage end of the distribution system to near power plants.

CHP systems, a form of DG, produce thermal energy and electricity from a single fuel source, thereby recycling normally wasted heat through co-generation (where heat and electricity are both used) and trigeneration (where heat, electricity, and cooling are produced). Recycled thermal energy is then used directly for air preheating, industrial processes that require large amounts of heat, space cooling, and refrigeration. As a result, CHP technologies consistently have double the efficiency of traditional utility power plants that vent heat into the environment as a waste product. The most significant market segments for CHP are industrial plants, district energy systems, and commercial buildings. For industries, CHP units can improve the efficiency of energy-intensive petroleum refining, the manufacturing of petrochemicals, and the paper and pulping industries. For district energy systems, CHP units can be configured to provide steam, hot water, and chilled water from a central point to a building or a network of buildings

Table 2.1 Renewable energy technologies and associated fuel cycles, 2012

Source	Description	Fuel
Onshore Wind	Wind turbines capture the kinetic energy of the air and convert it into electricity via a turbine and generator.	Wind
Offshore Wind	Offshore wind turbines operate in the same manner as onshore systems but are moored or stabilized to the ocean floor.	Wind
Solar PV	Solar photovoltaic cells convert sunlight into electrical energy through the use of semiconductor wafers.	Sunlight
Solar Thermal / Concentrated Solar Power	Solar thermal systems use mirrors and other reflective surfaces to concentrate solar radiation, utilizing the resulting high temperatures to produce steam that directly powers a turbine. The three most common generation technologies are parabolic troughs, power towers, and dish-engine systems.	Sunlight
Geothermal (conventional)	An electrical-grade geothermal system is one that can generate electricity by means of driving a turbine with geothermal fluids heated by the earth's crust.	Hydrothermal fluids heated by the earth's crust
Geothermal (advanced)	Deep geothermal generators utilize engineered reservoirs that have been created to extract heat from water while it comes into contact with hot rock, and returns to the surface through production wells.	Hydrothermal fluids heated by the earth's crust
Biomass (combustion)	Biomass generators combust biological material to produce electricity, sometimes gasifying it prior to combustion to increase efficiency.	Agricultural residues, wood chips, forest waste, energy crops
Biomass (digestion)	These biomass plants generate electricity from landfill gas and anaerobic digestion.	Municipal and industrial wastes and trash
Biomass (biofuels)	Liquid fuels manufactured from various feedstocks.	Corn, sugarcane, vegetable oil, and other cellulosic material
Hydroelectric	Hydroelectric dams impede the flow of water and regulate its flow to generate electricity.	Water
Ocean Power	Ocean, tidal, wave, and thermal power systems utilize the movement of ocean currents and heat of ocean waters to produce electricity.	Saline Water

through underground pipes. In Denmark, our case study below, most of the district heating system utilizes low temperatures – more than 90 percent of the total supply network operates with temperatures less than 100 degrees Celsius (212 degrees Fahrenheit).[23] For commercial buildings, CHP units can offer backup power at facilities such as hospitals or universities for when the power grid goes down or the price of grid-electricity suddenly rises.[24]

Renewable energy, another type of DG when deployed in smaller capacities, depends on nondepletable fuels that can be utilized through a variety of sources, approaches, systems and technologies shown in Table 2.1. Plants and algae can capture sunlight for photosynthesis before they convert it to biofuels or biopower. Hydropower capitalizes on the rain and snowfall resulting from water evaporation and transpiration. Wind generates electricity directly by turning a turbine or indirectly in the form of ocean waves, but the wind itself is driven by the sun. Tides go up and down due to the gravitational attraction between the oceans and the moon. The heat trapped in the earth itself can be put to productive use through geothermal applications.

The beauty is that renewable fuels are often indigenous and free; for the most part, any amount generated from sunlight or wind does not compete with the sunlight or wind available elsewhere. Countries need not expend considerable resources to secure renewable supplies. Put another way, a ton of coal or barrel of oil used by one community cannot be used by another, whereas renewable resources, because they are nondepletable, do not force such geopolitical tradeoffs. Moreover, the fuel cost for some renewables, such as wind and solar, can be known for decades into the future, something that cannot be said about conventional technologies, for which predicting spot prices in the future is about as accurate as reading crystal balls.

Furthermore, although renewable energy technologies have their own associated set of environmental and social impacts, they do not melt down, rely on hazardous fuels, or depend on a fuel cycle of mining or milling that must extract fuels out of the earth (with the exception of some biomass and landfill gas capture facilities). When roughly quantified and put into monetary terms, the negative externalities for coal power plants are 74 times greater than those for wind farms, and those from nuclear power plants are 12 times greater than for solar PV systems.[25] Oil, gas, coal, and uranium are susceptible to rapid escalations in price and price volatility, whereas renewable fuels are often free for the taking, widely available, and inexhaustible.

Case study: Danish energy policy

Denmark offers an excellent case study of a country that rapidly utilized energy efficiency, DG, CHP, and renewables to increase the availability of domestic energy and achieve self-sufficiency. From 1972 to today, its major objective has been to minimize dependence on foreign oil imports – something it has accomplished – and since then its chief goals have been maintaining energy independence, phasing out fossil fuels, and reducing carbon dioxide emissions.

At the core of Denmark's successful approach is a commitment to energy efficiency, prolonged taxes on energy fuels, electricity, and carbon dioxide, and incentives and subsidies for CHP and wind turbines. Major investments in energy efficiency mean that the country uses the same amount of energy today as it did in 1970, even though its economy and population have grown. Higher gasoline, diesel, and oil taxes were first passed in 1974 following the oil shocks from the Organization of the Petroleum Exporting Countries (OPEC) embargo and significantly expanded in 1985 when oil prices fell. These measures were followed by additional taxes for coal in 1982, carbon dioxide in 1992, and natural gas and sulfur in 1996. These taxes raised more than $25 billion from 1980 to 2005, and both helped Denmark avoid economic problems (such as inflation) associated with increasing exports of oil and natural gas, and created government revenue for energy efficiency and renewable energy research programs. CHP units powered by natural gas and biomass, including waste and straw, have expanded to supply a majority of the country's heat and a substantial amount of its electricity, and Denmark leads the world in both the per capita use of wind electricity and the share of wind electricity as a percentage of national supply.[26]

As Table 2.2 depicts, from 1980 to 2010 energy intensity in Denmark fell 27.8 percent, and gross energy consumption per capita declined 7.7 percent. The country's energy system was only five percent self-sufficient in 1980 but was 121 percent sufficient in 2010 (it had *extra* energy, though it was unable to successfully store all of it). Over this same period, the share of renewable energy in gross energy consumption has increased by 232 percent, wind turbine capacity has increased 638 percent, and CHP production has increased 61 percent (as a share of electricity) and 77 percent (as a share of district heating), while per capita carbon dioxide emissions have fallen 28.7 percent and emissions per unit of Gross National Product have declined 44.2 percent.

How did Danish policymakers achieve these feats? Their strategy rested on three central pillars (among others[27]): (1) their promotion of wind electricity, (2) their endorsement of combined heat and power and district heating, and (3) energy efficiency (inclusive of a tax on carbon and revenues funneled back into energy research).

History

Wind electricity

Denmark is endowed with some of the best wind conditions in Europe, convincing policymakers, in tandem with a global oil crisis, to embark on a government-sponsored wind research program in the 1970s.[28] Starting in 1979, the Danish government promoted an investment subsidy that

Table 2.2 Key energy statistics for Denmark, 1980–2010

	1980	1990	1995	2000	2005	2010
Energy intensity, gross energy consumption (TJ per DKK Million GDP, 2000 Prices)	0.998	0.818	0.748	0.649	0.618	0.591
Gross energy consumption per capita (GJ)	159	160	161	157	157	147
Degree of self-sufficiency (%)	5	52	78	139	155	121
Renewable energy – share of gross energy consumption (%)	2.9	6.1	7.0	9.8	14.7	20.2
Wind turbine capacity – share of total electricity capacity (%)	–	3.8	5.7	19.0	23.9	27.7
CHP production – share of total thermal electricity production (%)	18	37	40	56	64	61
CHP production – share of total district heating production (%)	39	59	74	82	82	77
Renewable energy – share of total domestic electricity supply (%)	0.0	2.0	5.9	15.3	17.8	33.1
CO_2 emissions per capita (tonnes)	12.2	11.9	11.5	10.4	9.7	8.5
CO_2 emissions per kWh electricity sold (gram CO_2 per kWh)	1 034	937	807	634	538	505
CO_2 emissions per GNP (tonnes per Million GDP)	77	61	53	43	38	34

Source: Danish Energy Authority.

reimbursed individuals, municipalities, and farming communities for the capital cost of installing wind, solar, and biogas digesters. These subsidies initially covered 30 percent of the expense of renewable energy systems. Although they were later scaled down periodically for wind turbines as the industry matured and prices decreased, Danish policy enshrined three important principles: (1) all farmers and rural households had the chance to install a wind turbine on their own land; (2) local residents had the possibility to become members of local cooperatives in their municipalities or neighboring municipalities, and exclusive local ownership was a condition for operating permits; and (3) electric utilities could only build large wind farms in agreement with the government and if this did not violate the wishes of farmers and local residents.[29] This "cooperative" approach to wind energy reflects a long history in the country of promoting cooperatives in agriculture and other sectors of the economy.[30]

Just two years later, in 1981, the government passed a feed-in tariff (FIT) requiring utilities to buy all power produced from renewable energy technologies at a rate above the wholesale price of electricity in a given distribution area.[31] In 1985, an agreement was reached between the government and the electricity utilities, committing the utilities to install 100 MW of wind capacity over a five-year period (which was fully implemented by 1992). That same year, lawmakers passed two other important policies. The government established the Danish Wind Turbine Guarantee. This guarantee provided long-term financing of large wind projects that used Danish-made wind turbines, encouraging local manufacturing and reducing the risk of building larger projects. The Danish Energy Authority also provided open and guaranteed access to the grid. Grid connection costs were to be shared between the owner of the wind turbine and the electricity utility. The wind turbine owner had to bear the costs of the low-voltage transformer and connection to the nearest connection point on the 10/20 kV distribution grid. The utilities had to cover the costs for reinforcement of the grid when needed.

As a consequence, Danish electricity distribution companies were legally obligated to finance, construct, and operate the transformer stations and transmission and distribution (T&D) infrastructure for centralized wind farms and decentralized wind turbines owned by ordinary people. They were obligated to connect wind power and expand the grid if necessary, and required to provide financial compensation if any of the wind power generated was curtailed. The costs of this infrastructure investment and reimbursement for curtailed power were paid for by electricity companies, and then distributed to all customers. The distribution company had the right to reject the connection if it could prove that the costs would become excessive. Rather than give up, however, in those rare cases the utility was required to present alternative solutions.

To accommodate the technical difficulty of managing a highly dispersed and decentralized electricity system, the Danish T&D network is comprised of mostly newer equipment with a preponderance of shorter transmission lines delivering power over smaller distances with low-voltage levels. A significant number of transmission lines are underground, and system operators report no unaccounted loss – in other words, the grid is so tight that they can "count every kWh." This tightness is needed to trade power between its Nordic neighbors in the north and Germany in the south, and it is bolstered by the large hydroelectric and pumped hydroelectric reserves of Norway and Sweden.

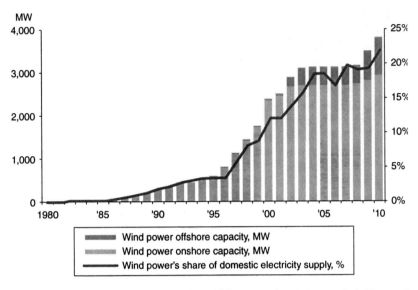

Figure 2.2   Wind power capacity and share of domestic electricity supply in Denmark, 1980–2010

Source: Danish Energy Authority.

Under this cooperative system, wind power grew from a mere 11 TJ of gross generation in 1978 to 28,114 TJ in 2010, and in the same year wind turbines supplied 21.9 percent of the country's electricity – numbers reflected in Figure 2.2.

District heating and CHP

To minimize dependence on fossil fuels and improve the efficiency of their fleet of power plants, Danish policymakers aggressively encouraged the use of CHP units for both electricity generation and district heating. From 1955 to 1974, almost all heating in Denmark was provided by fuel oil, which meant the oil crisis had particularly painful impacts on the country's economy.[32] In 1973, when oil prices rose astronomically, 85 percent of Denmark's electricity came from oil, its transport sector was almost entirely dependent on it, and oil provided more than 90 percent of the nation's primary energy supply.[33]

The Danish Energy Policy of 1976 therefore articulated the short-term goal of reducing oil dependence, and it stated the importance also of building a "diversified supply system" and meeting two-thirds of total heat consumption with "collective heat supply" by 2002. Moreover, it sought to reduce oil dependence to 20 percent, an ambitious goal that involved the need to convert roughly 800,000 individual oil boilers to switch to natural gas and coal.[34] In a mere five years – from 1976 to 1981 – Danish electricity production changed from 90 percent oil-based to 95 percent coal-based.[35] Stipulations in favor of CHP were further strengthened by the 1979 Heat Supply Act, whose purpose was to "promote the best national economic use of energy for heated buildings and supplying them with hot water and to reduce the country's dependence on mineral oil."[36]

To achieve the stated purposes of these two major acts, a "radical restructuring" of the heat supply system commenced, converting practically all oil-based systems to something combusting coal, natural gas, or biomass.[37] In undertaking this campaign, Parliament delegated authority away from Copenhagen to local municipalities, and emphasized the viability of both natural gas and biomass for CHP systems. In 1979, for example, the Danish Parliament established a national natural gas project based on offshore gas fields in the North Sea, and in 1981 district heating systems utilizing straw were introduced, leading to "serious straw-for-energy expansion."[38] In 1986, the Danish Energy Agency, Steering Group for Renewable Energy, and the Danish Board of Technology encouraged even more decentralized CHP generation and built straw demonstration plants ranging from 100 to 3,000 kW. All in all, Denmark invested about $15 billion from 1975 to 1988 in CHP systems and transmission networks, to the point where more than 350 district heating companies operated nationwide by 1990.

The 1990s saw continued support for CHP units. In 1990 a "triple tariff" system was introduced which paid CHP operators based on their provision of peak, medium, or low-load electricity and also granted them an energy "premium" of an extra 1.3 €¢ per kWh. At the same time, the Danish Parliament placed a moratorium on coal use and announced that "no new coal-fired power plants would be permitted," a proclamation later formalized in 1997 when the Danish parliament passed the "coal stop," functionally outlawing the construction of new coal-fired power stations, with exceptions given only to two 450 MW plants. Coupled with these improved tariffs and the moratorium on coal, the government promoted environmentally friendly zoning to advance electricity investments in towns and villages outside major cities. Co-generation units were required to replace district heating units, and their previous use of oil, diesel, and coal was prohibited and replaced by natural gas.[39] If the local market was not large enough to cater for co-generation, the district heating plants were required to utilize biomass. Concomitantly, all large utilities in the major cities were ordered to use biomass (especially straw), and required to obey mandatory energy efficiency regulations.[40]

This wave of environmentally friendly conversions and efficiency improvements drove significant investment in the CHP market. From 1990 to 1997, more than three-quarters of all new capacity added to the Danish grid consisted of small CHP plants for district heating or industrial use, fueled by natural gas or straw, to the point where today more than 40 percent of CHP fuels come from renewable resources and 25 percent come from natural gas.[41] From 1972 to 2010, CHP use in aggregate has expanded by a factor of four

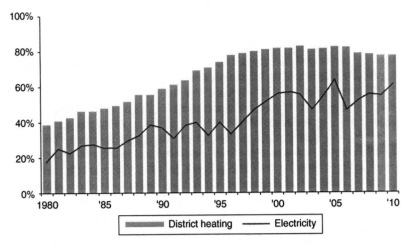

Figure 2.3   CHP share of electricity and district heating production in Denmark, 1980–2010

Source: Danish Energy Authority.

from 49,196 TJ to 200,870 TJ. As Figure 2.3 reveals, co-generation provides more than 50 percent of all electricity and almost 80 percent of district heat consumed in Denmark through approximately 45,000 kilometers of pipes,[42] making it the leader in Europe.[43] The next closest country, the Netherlands, harnesses only 38 percent for electricity; Finland, 36 percent; and every other country in the European Union less than 10 percent.[44]

Energy efficiency

Energy efficiency has been the third pillar of Danish energy policy. After the oil crisis of 1973, the government sponsored a public information campaign and tightened building codes so that annual space heating consumption was limited to less than 90 kWh per square meter. Parliament subsidized energy audits with public funds so that standardized reports of measures to reduce heat consumption could be attained, and it also initiated a large subsidy scheme to weatherize houses and improve the thermal insulation of residential and commercial buildings.[45] Since then, Denmark has aligned itself closely with the somewhat aggressive energy-efficiency policies and goals of the European Union (EU), especially given that the EU commissioner in charge of energy-efficiency regulation, Connie Hedegaard, was a former Danish Minister. Denmark's energy-efficiency efforts have fallen into four key areas: energy taxes and quotas, obligations, labeling, and the creation of an Electricity Savings Trust.[46]

Energy and carbon taxes   Denmark introduced an energy tax in 1977 for all households and a carbon dioxide tax in 1996 across all sectors. Regulators kept these energy-related taxes high after fossil fuel prices dropped in the 1980s and 1990s so that the development of a renewable energy industry could rely on predictable fuel and electricity prices. The taxes furthermore sent price signals that encouraged energy-efficiency measures in the Danish power market, and accrued government funds for R&D expenditures, which they directed to wind power, biomass, and small-scale CHP units. As a result of these taxes, primary energy consumption nationally has grown just four percent from 1980 to 2004, even though the economy grew more than 64 percent in fixed prices. The carbon tax alone added about 1.3 €¢ per kWh of additional income for renewable power providers. Between 1980 and 2005, these taxes raised roughly 330 billion Danish Krone (or $57 billion) in revenue.

These taxes keep consumption low – without them Danish energy consumption would be at least 10 percent higher – and kept fossil fuel prices high. Such high fossil fuel prices served to justify a government program prompted by researchers and local communities interested in using low-tech windmill designs to generate electricity. Danish regulators, working with manufacturers and interested citizens, took a bottom-up strategy of wind turbine development – a slow, crafts-oriented, step-by-step process including incremental learning through practical experience.[47] The Danish model adhered to "learning by doing," and designers recognized the learning curve needed to perfect development and were willing to tolerate and learn from earlier setbacks.[48]

Their approach paid dividends. According to the Risø National Laboratory in Denmark, from 1980 to 2005 the cost per kWh of Danish wind turbines decreased 60 to 70 percent, and Danish R&D enabled wind turbines to produce 180 times more electricity at 20 percent of the cost, due largely to improved capacity factors, or the amount of time a turbine can actually generate electricity. Over the same period, commercial turbine output grew 100-fold, from 30 kW in the 1980s to 3.0 MW in 2006. Based on such results, the government has actually *increased* energy taxes by 161 percent from 1990 to 2009.

Energy efficiency obligations   Denmark established an energy-efficiency "obligation" for electric utilities, mandating as of 2006 that all electricity, natural gas, and district heating providers realize energy-efficiency goals, efforts funded by a slight increase in energy bills that adds up to roughly €40 million per year. The "obligation" committed these companies to achieve specific demand-side management targets, and to contribute to an informational campaign on energy efficiency.

Energy labeling   Denmark regulated the use of energy labeling for buildings and appliances. For buildings, a labeling regime was established in 1979 (most recently updated in 2006) that consists of a label "A" to "G" with recommendations on how to reduce energy consumption. It also requires that a building must have an Energy Performance Certificate (EPC) specifying its energy usage whenever it is rented or sold. Furthermore, owners and real estate agents must label all new buildings before they enter use and existing buildings must update their label, which costs €650, whenever they are sold. Particularly large buildings (greater than 1,000 square meters in floor space) must update their labels every five years. The requirements from 2006 mandated a 25 percent reduction in energy use across all buildings by 2010 and a further 25 percent reduction in 2015 and 2020, respectively. Labels for domestic appliances in the home, such as televisions, refrigerators, and washing machines, used a similar lettering scheme and are estimated to save about 700 TWh of energy every ten years, and more than 90 percent of new appliances bought in Denmark have achieved the "A" efficiency rating.[49] This created also a new service industry of highly trained energy consultants who measure and rate buildings.

Electricity Savings Trust   Denmark created an Electricity Savings Trust from 1997 to 2010 to promote efficiency for homes and public buildings.[50] Initially tasked with incentivizing district heating, it started contributing towards the energy labeling regimes described above as well as increasing public awareness through information campaigns and counseling for households and companies until it converted into a Danish Energy Savings Trust in March 2010.[51] The Trust was funded by a €0.10 fee per kWh paid by all electricity customers. Some of its campaigns emphasized low-energy windows with special coatings; others sponsored a personal energy audit subsidized by the government.[52]

Recent developments

Though these three pillars of Danish energy policy have faced some challenges, discussed below, government plans have called for an even more determined commitment to renewable energy, CHP, and energy efficiency. In 2006 the Danish Prime Minister Anders Fogh Rasmussen announced a long-term target of "100 percent independency of fossil fuels and nuclear power," later presented as a national "Energy Plan of 2006."[53] This ambitious target was formalized into a Danish Society of Engineers (IDA) energy strategy called the "IDA Energy Plan," consisting of four prongs:

- Reduce energy demand in the long term, including a reduction in space heating demand in buildings by 50 percent, fuel consumption in industry by 40 percent, and electricity demand by 50 percent;

- Improve energy efficiency by encouraging heat pumps and solar thermal water heaters for homes and fuel switching for CHP units away from gas, coal, and oil;
- Expand renewable energy so that 30 percent of national supply comes from renewables by 2025 and 100 percent by 2050 (more specific targets include doubling the amount of wind capacity and introducing 500 MW of wave power and 700 MW of solar PV power);
- Promote intelligent energy systems that can better balance supply and demand, reduce transmission losses, and utilize "smart grid" technologies.[54]

If Denmark meets these IDA Energy Plan targets, by 2050 primary energy supply will fall significantly and carbon dioxide emissions will equal zero. Though it may sound unrealistic, one independent assessment concluded that for Denmark "a 100% renewable energy supply based on domestic resources is physically possible."[55]

Benefits

The social and environmental benefits from Denmark's energy taxes and support for energy efficiency, wind energy, and CHP have been impressive, and this section shows they have (1) lowered energy intensity, (2) realized national energy self-sufficiency, (3) drastically reduced greenhouse gas emissions, (4) prompted lucrative exports of Danish energy technologies, and (5) improved national energy security.

Lowered energy intensity

Danish investments in energy efficiency have considerably lowered its energy intensity, the amount of energy needed to produce a unit of GDP. From 1990 to 2008, the energy efficiency of final energy consumers in Denmark improved by 18 percent and primary energy intensity has declined by 26.3 percent, even though over the same period Denmark's economy has grown by 44.5 percent.[56] Final heat consumption in households declined ten percent from 1980 to 1999 while heated floor space actually increased by 19 percent.[57] Electric utilities also surpassed their energy-efficiency "obligation" targets by 25 percent in 2008. As a result, one independent study from the Regulatory Assistance Project, a think tank, of 11 industrialized countries concluded that Denmark's "pioneering efforts in building regulatory requirements for new and existing buildings, and the introduction of energy performance certificates for buildings, have been exemplary."[58]

Achievement of self-sufficiency

Denmark is now energy self-sufficient. Since 1991, the country has been self-sufficient in oil and gas. As Figure 2.4 reveals, the country transitioned

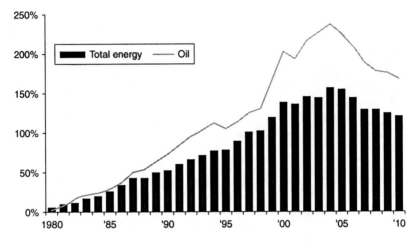

Figure 2.4   Degree of energy self-sufficiency in Denmark, 1980–2010
Source: Danish Energy Authority.

from being dependent on foreign sources for more than 95 percent of its energy in 1980 to becoming self-sufficient in 1997 and having an *excess* of energy every year after – the only country to do so in the entire European Union other than Estonia (which utilizes its cache of abundant fossil fuels). Denmark has an energy trade *surplus* and its energy patterns positively reflect on its balance of payments.

Fewer greenhouse gas emissions

As more renewable energy and co-generation units have come to replace less efficient and more polluting conventional fossil fuel units, related CO_2 emissions have plummeted. In 1990, almost 1 kilogram of carbon dioxide was emitted for every kWh of electricity produced in Denmark. In 2005, less than 600 grams was emitted. Altogether, the carbon dioxide emission intensity – the amount of CO_2 emitted per unit of GDP – was 48 percent lower in 2004 than it was in 1980. Denmark is also one of the few countries to have reduced absolute carbon dioxide emissions 19 percent below 1990 levels by 2009, surpassing its obligations under the Kyoto Protocol.[59]

Enhanced competiveness

Denmark used its expertise in wind turbine operation and manufacturing to create a robust market for export. Danish regulators promoted the localization of wind energy manufacturing by offering a strong home market for wind technology and stable annual demand. As such, they created opportunities through the sales of new products, jobs, and an increased tax base. They also used their domestic turbines as a real world laboratory to experiment,

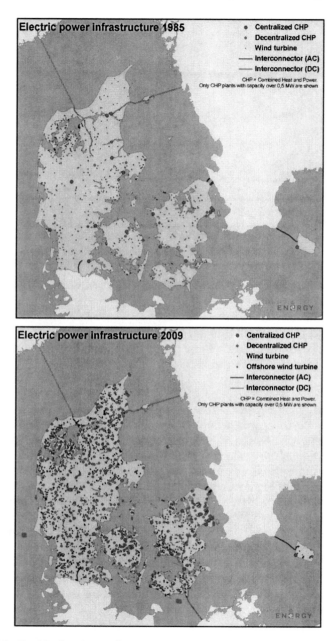

Figure 2.5 Danish electricity infrastructure in 1985 and 2009
Source: Danish Energy Authority.

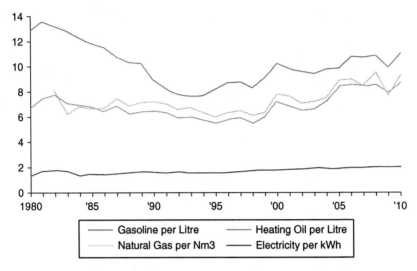

Figure 2.6   Energy prices for Danish households (DKK $2010), 1980–2010
Source: Danish Energy Authority.

lower the cost of turbine equipment, and improve capacity factors, though the sector has started to face fierce competition from overseas wind energy firms, especially in China and India.[60]

Improved energy security

Lastly, the greater penetration of co-generation and wind energy has promoted the decentralization of energy supply, a transition that has brought political benefits (in the form of greater diversification) and energy security benefits (an electricity system with a high fraction of distributed wind has more power plants on the grid, but also fewer individual units whose capacity is large enough to destabilize the system in the event of an accident or terrorist attack). The country shifted from centralized generation (with fewer than 20 large-scale plants) in the 1970s to a decentralized model including more than 4,000 small-scale generators today, as shown in Figure 2.5. Danish perceptions of wind energy are thus very positive,[61] and the country has seen its energy prices (adjusted to $2010), as shown in Figure 2.6, remain remarkably stable. One international comparison of the energy security performance of 22 countries within the OECD concluded that Denmark was the *most* energy secure.[62]

Challenges

Notwithstanding the merits of Denmark's approach to energy policy, it has faced some challenges related to (1) consistency, (2) managing a dispersed network, (3) bottlenecks, (4) social opposition, (5) rising prices, and (6) a potential return to fossil fuel imports.⌋

Changes in government policy

One substantial challenge has been a switch in government policy that started in the late 1990s, intending to liberalize and restructure the Danish energy sector to make it more market-based and competitive. In 1998, the government abolished its principles encouraging local and cooperative ownership of wind farms, removing restrictions that enabled electric utilities to consolidate operations and cross-invest in energy infrastructure. In 2001, a liberal conservative government took over and there was a meaningful change in Parliament. The result was an alteration of Danish energy policy by promoting a "sanitation program" to "improve the landscape" by removing wind farms from it. The idea behind this program was that a preferred course of action was to have a small number of very large turbines instead of a large number of very small ones. That year, the tariff for wind turbines was also set at the wholesale market price plus a small subsidy (functionally replacing their earlier FITs), and in 2004 the government abolished purchasing obligations for utilities beyond the minimum amount of full load hours.[63]

Wind developers, however, and even those within the Danish Energy Authority, appear to concur that these policies stunted wind growth. From 2004 to 2006, developers installed less than 40 MW of capacity, compared with more than 1,000 MW before the changes were made. In 2006 a meager 8 MW was installed annually and in 2007, for the first time ever, more wind capacity was decommissioned (39 turbines) than installed (seven turbines) and the number of windmill owners and co-owners dropped from a few hundred thousand to less than 50,000. During this time, DONG Energy and E.ON, two major developers, also switched their emphasis to offshore wind parks at Nysted/Rodsand II and Horns Rev II. In 2008, Danish policymakers seemed to realize their mistake and recommitted themselves to wind energy, but the period of 2001 to 2007 was certainly "stagnant."[64] Major factories in the wind sector also closed during this period, negatively impacting rural economies.

Managing a dispersed network

The decentralized nature of wind and distributed generation units has improved energy security, but it has also made managing a dispersed transmission and distribution grid network more difficult. Unlike many other countries, Denmark does not have a single, integrated national grid, but two historically distinct transmission areas – Eltra in the West, and Elkraft in the East, merged together in 2005 into Energinet Danmark – unable to fully assist each other in power balancing since they share only a single

DC connection.[65] Though the inclusion of CHP and wind into the Danish portfolio has not degraded overall reliability (and has attained the energy security benefits described above),[66] excess capacity in both regions must sometimes be "dumped," and the ability to "store" wind energy in Germany and Sweden will decline as those countries bring their own wind turbines online in the future,[67] though Norwegian planners have stated they will try to maintain their storage capacity, given that it provides them the ability to purchase "dumped" energy and resell it at a premium.

The decentralized Danish electricity network is not only difficult to manage technically, but in particular places difficult to manage socially and politically. The hundreds of cooperatives and independent CHP providers ensure a large diversity of perspectives, but they can fragment and complicate energy decision-making. One independent study warned that in Denmark "there is a strong need for better coordination of municipal energy planning activities at the central level ... the role of municipalities as energy planning authorities needs to be outlined more clearly" and concluded that Denmark needs to provide these dispersed actors with better planning instruments so that their efforts can be harmonized.[68]

Bottlenecks, lack of skills, and outsourcing

The rapid growth in the Danish wind industry and expansion to meet not only domestic demand but also exports for a strong global market has induced bottlenecks in manufacturing, construction, and wind farm design. The average wait for wind turbines from Vestas is now between 24 and 30 months; LM Glasfiber, a manufacturer of turbine blades, reports a minimum two-year delay on their products. For offshore wind turbines, 12 to 18-month delays are common for cables and 18 to 36-month delays are typical for installation vessels, which the oil and gas industry also rely on to install platforms and drills (an indication that wind projects must compete with profitable petroleum projects around the world). Bottlenecks have arisen in offshore wind farm construction, due mainly to a shortage of vessels capable of mounting turbines to the seafloor as well as suitable harbors where offshore wind turbines can be assembled. Part of the explanation is the booming Chinese market; China consumes a third of the total steel produced in the world and leads the globe in wind energy installations, and its market share has tripled in just ten years. That has pushed up the world price for raw materials, provoking a significant price increase for wind turbines. It has also motivated manufacturers in Denmark, such as Vestas and Siemens (which purchased the Danish wind energy company Bonus in 2004), to outsource the construction of wind turbines and components to facilities in India and China, closer to those emerging markets, and with lower labor costs.[69]

Rising social opposition

Though Danish attitudes towards wind energy remain mostly positive compared with other countries,[70] larger wind turbines and consolidation of wind farms among corporate actors have started to engender an increase in social protest and opposition, especially over bigger turbines that obstruct environmental viewscapes and exclude local cooperatives and farmers from ownership and participation. As a result, one study found an "added pressure" against wind turbines for the reasons that they have become too large and "visible" for some Danes to approve.[71] A second inquiry into attitudes towards wind power in Northern Jutland cautioned that "ever-increasing turbine size and less local involvement ... could ultimately lead to a decline in the popular acceptance of wind power."[72]

Price volatility

Another challenge concerns price volatility. Given the laws of supply and demand, wind power offers a low market price precisely when its output is high (because it is more likely that many turbines are generating excess power), meaning electricity is also valued the least. Conversely, wind output may be low when power is needed. This type of price volatility occurs both hourly and monthly. In January 2005, the amount of available wind capacity in Denmark dropped below 100 MW as turbines shut down during a hurricane, forcing system operators to switch on expensive peaking plants and to increase imports from Germany and neighboring Nordic countries.

Return to fossil fuels

A final challenge connects to Denmark's overall energy policy. While definitely supportive of renewables, the Danish energy sector is still dominated by fossil fuels. Although the Danish government has publicly iterated a plan to be 100 percent independent from reliance on fossil fuels, it still consumed 7.2 million short tons of coal in 2010 and produced 223,480 barrels of oil per day throughout 2011.[73] Of the total energy actually produced and used in Denmark, 40 percent came from oil, 23 percent from coal, 30 percent from natural gas, and only 17 percent from renewable resources.[74] Because of this mix, Denmark is still "one of the nations in the world with the highest level of carbon dioxide emissions per capita."[75] Wind and CHP units help improve Denmark's performance in the electricity sector, but its transport sector is "entirely fossil fuel based"[76] and analysts have argued that transitioning away from petroleum-based transport fuels will be hard for the country to achieve, given its climate and existing infrastructure.[77] As Figure 2.7 depicts, Denmark's transport, agricultural, industrial, and even household sectors remain both carbon-intensive and fossil-fueled.

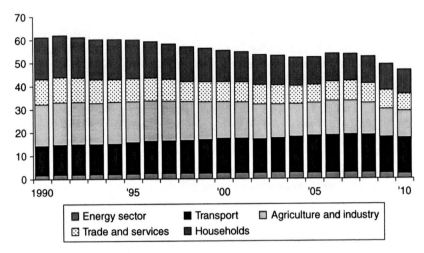

Figure 2.7   CO$_2$ emissions from final energy consumption in Denmark (million tons, climate-adjusted), 1990–2010

Source: Danish Energy Authority.

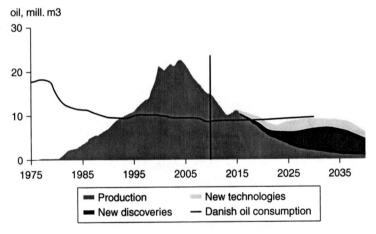

Figure 2.8   Oil production in Denmark, 1975–2040

Source: Danish Energy Authority.

Moreover, Danish oil and gas reserves are rapidly running out. Danish oil reserves have recorded an average annual decline of 6.7 percent, and under these conditions the country is expected to become an importer of oil and gas (again) by 2018, as Figure 2.8 illustrates.[78] As energy professor Henrik Lund recently commented, "Danish oil and gas resources are scarce...An interesting question is therefore can Denmark convert to 100 percent renewable energy within a matter of decades or will we have to return once again to former days of dependency on import of fossil fuels?"[79]

Conclusion: lessons and implications

Still, the Danish approach to energy policy offers at least five important lessons for other countries.

First, it empirically proves that national energy transitions are possible, and that they can occur quickly. After the international oil crisis of 1973, it took Denmark only five years to switch from being 95 percent dependent on oil for electricity generation to five percent. In roughly two decades, Denmark transitioned from a centralized electricity network based on large-scale fossil-fueled plants to one featuring wind turbines and CHP units, to the degree where Denmark is still the world leader in wind power (as a percentage of its portfolio and on a per capita basis) and the leader among the EU in its use of CHP.[80]

Second, the Danish experience proves that a carbon tax does not have deleterious effects on the overall economy, and that, if implemented properly, it can be a useful tool for promoting wind energy, energy efficiency, and CHP. Moreover, even though Denmark has a number of expensive energy taxes, its economy has still grown at double digit rates over the past few decades.

Third, guaranteed and open access to the grid exhibits numerous advantages. It minimizes barriers to market entry and prevents utilities from using their power of incumbency to block renewable energy projects on transmission and distribution grounds. It increases the profitability of renewable energy projects by shifting the costs of interconnection to the grid from the project developer to the utility (and the consumer, who ultimately pays for it). It also encourages community ownership which leads to higher rural incomes and greater social acceptance.

Fourth, Denmark validates a "polycentric" approach to energy planning which emphasizes engaging stakeholders at multiple geographic scales. Each of the main tracks of the Danish approach – wind, CHP, and energy efficiency – divided responsibilities among both national and local actors. For wind power, national planners offered stable financial support and developed appropriate guidelines, while local planners and cooperatives drafted wind power plans and supported specific wind projects. For CHP, national planners provided appropriate tariffs and clear guidelines about minimizing the use of oil and coal, while local planners carried out heat plans and accelerated the connection of buildings to district heating. For energy efficiency, national planners regulated things like building codes and the National Electricity Trust, while local actors built partnerships between banks, utilities, and residents.[81] These efforts embraced local participation, and as a result the ownership of Danish energy projects remains decentralized at the local level rather than concentrated in the hands of large corporations. In 2005, for example, only 12 percent of wind farms were utility-owned; individuals and cooperatives owned the remaining 88 percent.[82]

Fifth, and finally, is the necessity of consistency. Danish energy policy was remarkably consistent from 1973 to 1998, and that stability saw its markets for wind energy, CHP, and energy efficiency expand dramatically. The inverse holds true as well, with the changes implemented from 1998 to 2007 essentially derailing investments in wind and causing the industry to stagnate unnecessarily. Luckily for Denmark, its political system changed again in 2008 and a more liberal government has placed the country on a pathway to freedom from fossil fuels by 2050 – a testament to both a very strong and active energy lobby, and the realization that sometimes energy transitions can occur only under the "right" political circumstances.

3

Affordability and Fuel Poverty in England

Introduction

Having a warm, comfortable, well-lit home is an energy justice ideal. Without sufficient warmth or electricity, modern families must suffer the harsh climates of winter without heat, leading to hundreds of thousands of excess winter deaths every year, or rely on "coping" strategies such as cutting down on expenses related to food or medical care to pay their energy bills. After presenting the issue of affordability and fuel poverty as a justice concern, the chapter articulates the connections between thermal comfort, mental health, and physical health before detailing the frequently inequitable nature of household energy consumption, with poorer households paying a greater share of their income on energy services than wealthier and middle-class households. It finishes by analyzing the case of England's Warm Front Home Energy Efficiency Scheme, an effort intended as part of the Warm Homes and Energy Conservation Act of 2000 to eradicate fuel poverty throughout the country by 2016. Though it remains unlikely the Scheme will accomplish that ambitious target, from 2001 to 2011 it has removed almost 2.3 million households from fuel poverty. Warm Front interventions have been credited with reducing carbon dioxide emissions per home by 1.5 tons per year, displacing £610.56 in modeled, potential annual energy costs and generating an average annual increase in income per customer of £1,894.79, and saving 11.2 GJ of energy per household each and every year for the next 20 years. Moreover, presuming the program's own reports are accurate, the Warm Front scheme has accomplished these tasks with an extremely high satisfaction rate, with an average annual customer satisfaction score of 92.3 percent for its most recent year.[1]

Affordability as an energy justice concern

Affordability matters from a justice perspective – when energy prices rise and households cannot afford them, it is functionally the same as if they lacked access to reliable energy services altogether. In addition, less affluent families spend a larger proportion of their income on energy services (12 percent of household income, for instance, for the poorest quintile of households in the United States), hindering the accumulation of wealth needed to make investments to escape their poverty. One study looking at the effects of increases in energy prices in four developing Asian economies from 2002 to 2005 found that poorer households paid 171 percent more of their income for cooking fuels and 120 percent more for transportation, 67 percent more for electricity, and 33 percent more for fertilizers when compared with the expenditures on energy by middle- and upper-class households.[2]

Environmental justice theorists Gordon Walker and Rosie Day have compellingly argued that affordability cuts across issues of both procedural justice and justice as recognition, a framing illustrated by Figure 3.1.[3] Fuel poverty impacts procedural justice because fuel-poor households have little time to participate in energy policymaking decisions or to learn about energy efficiency, and it impacts justice as recognition because being fuel-poor "can be read as a lack of recognition of the needs of certain groups, and, more fundamentally, as a lack of equal respect accorded to their wellbeing. This is particularly significant given that fuel poverty does not have equal consequences across different social and demographic groups."[4]

The concept of "fuel poverty" is thus one useful way of framing the issue of affordability. Initially defined as "households with high fuel expenditure as those spending more than twice the median on fuel, light and power," it is now commonly associated with households that must spend greater than 10 or 15 percent of their monthly income on energy services.[5] The European Union states that it is "persons, families and groups of persons whose resources (material, cultural and social) are so limited as to exclude them from the minimum acceptable way of life in the Member State to which they belong."[6] Conceptions of fuel poverty have been heavily influenced by Oxford University academic Brenda Boardman,[7] who initially argued that "fuel poverty occurs when a family is unable to afford adequate warmth because they live in an energy-inefficient home,"[8] that fuel poverty "occurs when a household is unable to afford adequate energy services in their home on their present income," and that it "relates to consistent, defined standards of energy services, not just actual expenditure."[9] Her work has been primarily concerned with fuel poverty in the United Kingdom, where she calculated that the poorest 30 percent of households spend less money per person on fuel than the other 70 percent of households, but pay twice as much as a proportion of their monthly budget. As Figure 3.2 reveals, using government data in 2006 she calculated that an astounding 68 percent of homes with incomes in the lowest decile in the United Kingdom were in fuel poverty.

The problem is most certainly not limited to the United Kingdom. In New Zealand, about one-quarter of homes suffer from fuel poverty due to a generally poor quality of housing in terms of thermal efficiency, comparatively

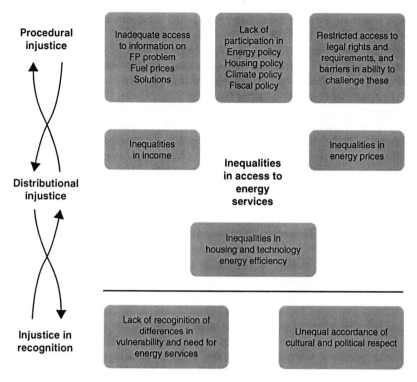

Figure 3.1 The energy justice implications of fuel poverty
Source: Adapted from Gordon Walker.

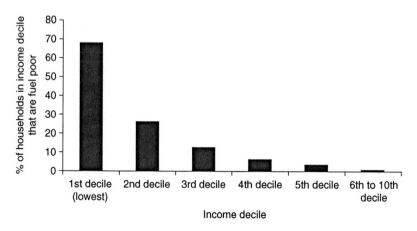

Figure 3.2 Households in fuel poverty by income decile in the United Kingdom, 2006
Source: Adapted from Brenda Boardman.

high levels of income inequality, and rapid increases in the real price of residential electricity.[10] In Austria, a "large number of households live in deprived conditions, carrying multiple burdens (lack of financial resources, energy-inefficient dwellings, old devices, energy costs or long-term illnesses)."[11] In Hungary, the problem of fuel poverty can be too *much* heat that cannot be controlled rather than too little. There, the fall of the Soviet Union "progressively brought energy prices to full-cost recovery levels, reduced household incomes and left a legacy of inefficient and deteriorating residential buildings lacking basic energy efficiency requirements."[12] When Hungarians cannot pay their heating bills, in serious cases they can lose their homes, as accumulated housing and utility bills force families to move to less valuable properties as a way to repay their energy debts.[13]

Perhaps an obvious point by now, poverty, influenced deeply by income, is not identical with fuel poverty, conditioned by household income as well as fuel prices and the energy efficiency of residential building stock.[14] As one study put it bluntly, "raising incomes can lift a household out of poverty, but rarely out of fuel poverty."[15] As another team of researchers have noted:

> Fuel poverty is a difficult concept. It is not the same as poverty. Some people are poor but can afford adequate warmth. Others with incomes above the accepted poverty line nevertheless cannot afford to be warm – because their home is difficult or expensive to heat. There are also people who purchase warmth only at the expense of adequate diets or going short in other ways. There are also those who live on in cold conditions despite having incomes which are sufficient to purchase adequate warmth – because of helplessness or fear of fuel bills.[16]

Thus, low-income households with sound investments in energy efficiency may not be in fuel poverty, whereas households with larger incomes in less efficient homes may be in fuel poverty.[17] Though the classic definition of fuel poverty is the "ten percent of household income" metric, another way of measuring it is those households that actually spend more on energy than on food.[18] Still others rely on the term "severe fuel poverty" to indicate needing to spend 15–20 percent of household incomes on energy, that is, between three and four times the median for a given year, and "extreme fuel poverty," those needing above 20 percent or greater than four times the median for a given year.[19]

The impacts of fuel poverty extend well beyond defaulting on energy bills. Mary O'Brien, Specialty Registrar in Public Health at the Wessex Deanery in the United Kingdom, has found that fuel poverty results in "inadequately heated housing" and, as a result, higher rates of mortality among the elderly, a greater prevalence of circulatory and respiratory diseases in adults, reduced physical and emotional wellbeing, and an increased risk of falls, mental health illness, social isolation, and hospital admissions.[20] More severely, fuel poverty quite literally kills people who go without essential heat and then suffer "excess winter mortalities." One epidemiological study looked at 11 industrialized countries in both the northern and southern hemispheres and found a clear correlation between the winter months and unusually high rates of mortality, a trend they plotted in Figure 3.2.[21] Shockingly, Table 3.1 shows the average excess winter deaths – defined as the extra deaths in the four winter months in comparison with the previous and succeeding four months – across a dozen countries and calculates that these amount to 278,409, exceeding the global number of deaths (about 166,000) attributed by the World Health Organization (WHO) to climate change. This makes fuel poverty as urgent a health issue as climate change, given that a 2006 WHO review of ten countries projected that the attributable fraction of excess winter deaths due to housing conditions was 40 percent.[22]

Such findings have been confirmed by scores of independent assessments.[24] To cite a few prominent examples, University of Ulster Psychology Professor Christine Liddell has found that cold-related deaths from fuel poverty can occur through "changes in blood pressure and blood chemistry during cold weather, which in turn increase the risk of catastrophic cardio- or cerebrovascular events such as strokes, myocardial infarctions or pulmonary embolisms" as well as the suppression of immune systems.[25] Research evidence examined by the Marmot Review Team demonstrated that countries with more energy-efficient housing had lower excess winter deaths,

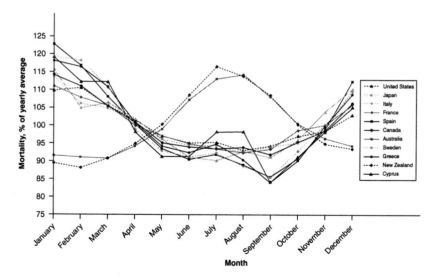

Figure 3.3 Monthly mortality for various developed countries as a percentage of yearly average

Source: World Health Organization.

Table 3.1   Average excess winter mortality in 12 countries, 2008[23]

Country	Population (millions, in 2008 when their study was done)	Average Excess Winter Mortality
Australia	19.5	6,973
Canada	30.3	8,113
Cyprus	0.8	317
France	59.5	24,938
Greece	9.5	5,820
Italy	54.4	37,498
Japan	127.9	50,887
New Zealand	3.5	1,600
Spain	37.5	23,645
Sweden	8.8	4,034
United Kingdom	60.9	36,700
United States	287.3	77,884
Total		278,409

Source: Adapted from Falagas *et al.* and Day and Hitchings.

that children living in cold homes are more than twice as likely to suffer from a variety of respiratory problems as children living in warm homes, and that cold housing increases the level of minor illnesses such as colds and flu and exacerbates existing conditions such as arthritis and rheumatism. It also noted that cold housing negatively affects dexterity and can heighten the risk of accidents and injuries in the home.[26] A team of researchers led by the WHO's Collaborating Center for Housing Standards and Health documented that "an inadequate supply of energy may also mean an inadequate supply for other basic domestic needs such as for food storage and cooking, maintenance of personal and domestic hygiene, and artificial lighting. Each of these could result in threats to health such as food poisoning, spread of infections, slips and fall injuries, fire injuries (from candles or oil lamps) and carbon monoxide poisoning (from inappropriate unflued heat sources)."[27] In another publication, the WHO noted that thermal comfort is "inextricably linked to health."[28] When it doesn't kill and sicken people directly, fuel poverty forces households to "cope" with inadequate energy services in a variety of ways. Surveys of the fuel-poor have found that they will resort to wearing coats and outdoor clothing indoors, sleeping together with pets or in one room to keep warm, having hot drinks, or even staying with relatives – actions that can all negatively impact mental health.[29]

Case study: fuel poverty in England

England's Warm Front Home Energy Efficiency Scheme, hereafter referred to as Warm Front (WF), successfully reduced the extent of fuel poverty among millions of English homes from 2001 to 2011. WF was "a major component of government strategy to eliminate fuel poverty in England and enable even the poorest households to maintain healthy indoor temperatures."[30] Managed initially by the Department for Environment, Food, and Rural Affairs, it provided grant-funded packages of insulation and heating improvements to eligible households. The scheme has lowered greenhouse gas emissions associated with low-income households, increased incomes and accrued billions of pounds of energy savings, and achieved these feats with strong approval ratings from its customers.[31]

History

The WF scheme began in June 2000 after a 1999 Inter-Ministerial Group on Fuel Poverty was instructed to develop a formal Fuel Poverty Strategy intended to identify and then address the causes of fuel poverty throughout all of the United Kingdom.[32] Based upon their input, the Warm Homes and Energy Conservation Act was passed in November 2000, culminating in the creation of the UK's 2001 Fuel Poverty Strategy to assist the fuel-poor primarily through WF in England and through other similar schemes in the devolved administrations of Scotland, Wales and Northern Ireland.[33] The 2001 Fuel Poverty Strategy officially noted that "a fuel poor household is one that cannot afford to keep adequately warm at reasonable cost. The most widely accepted definition of a fuel-poor household is one which needs to spend more than 10% of its income on all fuel use and to heat its home to an adequate standard of warmth. This is generally defined as 21 °C in the living room and 18 °C in the other occupied rooms."[34]

More specifically, WF provided packages of "insulation and heating measures depending upon the needs of the householder and the construction of the property." In 2001, it offered grants of up to £1,500 to those wishing to install loft and/or cavity wall insulation, draught proofing, gas wall heaters, dual element foam insulated immersion tanks, and heating repairs and replacements. This first grant was available to households with children or pregnant women on certain qualifying income-related benefits, as well as those in receipt of disability benefits. Under a "WF Plus" component, it offered grants of up to £2,500 to households claiming qualifying income-related benefits and occupied by those over the age of 60 for insulation measures and heating system upgrades, including new central heating systems.[35] The WF scheme was altered after the Government Spending Review in 2010. This resulted in a substantial reduction of the WF budget and a change in the eligibility criteria of the scheme, ostensibly to better target vulnerable households living in energy-inefficient homes. From 2011 to 2012 households had to be in receipt of income-related benefits *and* be living in a thermally inefficient home *and/or* not have a working central heating system. In its revised state, qualifying WF households could receive up to £3,500 in grants or £6,000 where oil central heating and other alternative technologies were recommended, though the entire scheme is scheduled to end in March 2013. Figure 3.4 shows the types of upgrades made under both the WF and WF Plus schemes.

The WF scheme works either by first having homeowners and tenants self-apply online, over the phone, and through the mail, or by referrals from previous customers. If an application meets WF's eligibility criteria, a WF surveyor visits the home to conduct a residential energy audit. During this audit, the surveyor will in some cases correct simple problems that require less than three hours of time; in other cases surveyors will rule out homes that are already energy-efficient. If homes pass inspection, they are allocated an installer who will suggest energy efficiency measures, prices, and an installation date. The final phase involves the actual installation of retrofits and equipment, with work supposed to be completed within three months for insulation and within five months for heating systems. After installation, the WF team inspects a random sample of five percent of homes with new heating systems and ten percent of those with insulation.

WF was strategically designed to be different from earlier home efficiency efforts. It originally had two scheme managers, Eaga Partnership and TXU Powergen, instead of relying on a single actor or a government agency.[36] Actual installations of energy efficiency equipment were allocated to installers through competitive bids placed electronically via email.

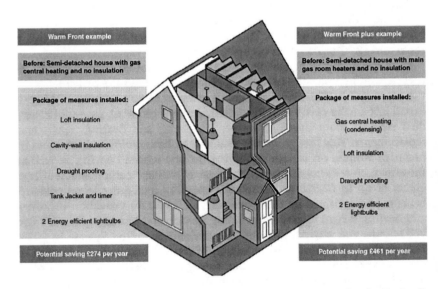

Figure 3.4   Common household energy efficiency upgrades sponsored under England's Warm Front Scheme, 2004

Source: National Audit Office.

The scheme provided training and "aftercare service visits" at least once a year so that customers received free service to "prolong the life of their systems."[37] Older schemes targeted almost exclusively those in public housing and offered a grant maximum of only £315, whereas WF applied to both private rented and owner-occupied housing and increased the minimum grant amount by almost five times. Older schemes permitted only insulation, whereas WF expanded eligible technologies to include insulation measures as well as heating systems, boiler replacements, energy audits, energy-efficient light bulbs, timer controls for space and water heating, and hot water thermal jackets. WF also relied on a "fuel poverty indicator" to predict the extent of fuel poverty in different household types, comprised by a national English House Condition Survey and Census data.[38] WF, lastly, was given more explicit and "reasonably practicable" targets to accomplish. It was required to (a) help 800,000 homes experiencing fuel poverty by 2004, (b) reduce household fuel use by 60 percent in participating homes, and (c) reduce cold-related deaths and related diseases by 2010. The Warm Homes and Energy Conservation Act, operating in tandem with WF, (a) set up a framework for completely eradicating fuel poverty country-wide in all vulnerable households (occupied by the disabled or long-term sick, elderly, or children) by 2010, and (b) set the goal of eliminating fuel poverty in all households by 2016.[39] Table 3.2 shows how WF started with an annual budget of roughly £150 million per year but expanded to a peak of £397 million in 2008–2009.[40] Figure 3.5 shows a WF brochure from 2005.

Table 3.2 Annual budget expenditures for England's Warm Front Scheme, 2001–2013

Fiscal Year	Budget (millions of £)
2001–2002	150
2002–2003	150
2003–2004	150
2004–2005	150
2005–2006	190
2006–2007	306
2007–2008	345
2008–2009	397
2009–2010	360
2010–2011	195
2011–2012	110
2012–2013	100
Total	2,603

Source: National Audit Office.

Figure 3.5   A Warm Front Scheme brochure from February, 2005
Source: Warm Front.

Benefits

WF did not accomplish all of its targets or those enshrined in the Warm
Homes and Energy Conservation Act, but it did result in a multitude of benefits related to (1) successful investments in household energy efficiency, (2)
thermal comfort and customer satisfaction, (3) improved health, and (4) a
net positive cost curve (meaning it saved more than it cost).

Rapid investments in household energy efficiency

The most obvious benefit was investments in household energy efficiency
measures that otherwise wouldn't have happened given the low incomes of

Table 3.3 Private installer prices compared with England's Warm Front Scheme prices

Technology	Average price based on private installers (£)	Average price for Warm Front Scheme participants (£)	Price savings (£)	Price savings (%)
New natural gas fired central heating with five radiators	3,463	2,325	1,138	33
New natural gas fired central heating with six radiators	4,790	3,951	839	18
Like-for-like gas fired boiler replacement	1,661	1,491	170	10
Like-for-like oil fired boiler replacement	2,455	2,196	259	11

Source: Adapted from National Audit Office.

the fuel-poor in the UK. In 2002, after its first year of operation, WF assisted 303,000 households, which received an average of £445 worth of energy efficiency improvements that reduced annual fuel bills by £150 per year.[41] By 2004, it reached more than 700,000 households, though the average household received only 1.1 energy efficiency measure.[42] Nonetheless, WF implemented these investments quickly and more cheaply than other private sector actors could have. An independent review from the National Audit Office estimated that it took an average of 64 working days to have a heating system installed and only 27 working days to have a property insulated – meaning that even major upgrades to homes were done in a matter of weeks rather than months.[43] Furthermore, it was able to do so at lower prices than private installers, figures reflected in Table 3.3, and more cheaply than other sources of electricity. Presuming that each household saves about 11.2 GJ per year for the next 20 years, the amount modeled and reported by the government, then the WF saves energy at a price of about 2.4 p/kWh, equivalent to 3.85 cents/kWh.[44]

Due in part to the WF scheme and in part to falling energy prices, fuel poverty dropped drastically throughout the UK from 5.1 million homes in 1996 to 1.2 million homes in 2003 during the first few years of the program. As of 2011, WF had reached 2.3 million homes (with 130,000 households receiving assistance that winter). Notable achievements include more than one million homes refitted with draught proofing and cavity wall insulation, 720,985 lofts insulated, the replacement of 454,828 boilers, and 73,154 new electric central heating systems installed, among others.[45]

Thermal comfort and customer satisfaction

Extensive, independent evaluations have revealed that the WF scheme has improved indoor living temperatures and received high rates of customer satisfaction. A major national health impact assessment of the WF scheme undertaken by the WF Study Group revealed that "the scheme has significantly raised average indoor temperatures."[46] A series of household surveys, physical monitoring, and energy audits with thousands of homes confirmed significant increases in indoor temperature among participants. Moreover, the study found that WF increased daytime living room temperatures by an average of 0.58 to 2.83 degrees Celsius, that it reduced the amount of energy needed to maintain warmth by five to ten percent, and that it improved air quality in the home, with less detectable dust and mold.[47] WF also increased the proportion of households reporting being "thermally comfortable" from 36.4 percent to 78.7 percent.[48] This large national assessment concluded that "WF energy efficiency improvements lead to substantial improvements of both living room and bedroom temperatures which are likely to have benefits in terms of thermal comfort and wellbeing."[49]

Also noteworthy is the high level of self-reported satisfaction with the WF scheme. A second independent evaluation from the National Audit Office in 2009 found that 86 percent of households assisted by the Scheme were "satisfied with the quality of work done," and only five percent were "dissatisfied." Moreover, the most common complaint among this five percent related to delays, implying that the problem was not with energy efficiency equipment *per se* but with not receiving it quickly enough.[50]

Improved public health

The WF scheme improved the physical and mental health of England's fuel-poor. The fuel-poor in England are more prone to excess winter deaths caused by exposure to cold and the incidence of circulatory and respiratory disease, found to be three times higher in the coldest quarter of housing stock compared with the warmest 25 percent.[51] Fuel poverty in the UK has been correlated with higher rates of hospital admission, as well as reductions in the budgets for food and other basic commodities, leading to "a fall in calorie intake for adults and children at a time of increased need for nutritional energy" as households "cope" by cutting back on essentials to pay their fuel bills.[52] For example, the household survey depicted in Table 3.4 shows that almost two-thirds of fuel-poor homes either "turn off their heat" or "turn it down" to save money.[53]

The WF Scheme, in response, reduced morbidity and mortality associated with fuel poverty and has also lessened the risk of chronic medical

Table 3.4 Coping strategies undertaken by fuel-poor households in the UK

	% households surveyed
Turned heating off, even though would have preferred to have it on	35%
Turned the heating down, even though would have preferred it to be warmer	33%
Turned out lights in my home, even though would have preferred to have them on	22%
Turned the heating down or off in some rooms but not others, even though would have preferred not to	20%
Only heated and used one room in the house for periods of the day	14%
Used less hot water than would have preferred	15%
Had fewer hot meals or hot drinks than would have liked	4%
None	37%

Source: Adapted from Anderson *et al.*

conditions such as cardiovascular and cerebrovascular diseases, diabetes, respiratory and renal diseases, Parkinson's disease, Alzheimer's disease, and epilepsy.[54] The Warm Front Study Group found "significant health benefits" not previously estimated, advantages including higher temperatures, satisfaction with their heating system, greater thermal comfort, and less stress. They concluded that WF resulted in the "alleviation of fuel poverty and the reduction of stress associated with greater financial security" and that it was "more successful than implied."[55] Many households also reported "perceptions of improved physical health and comfort, especially of mental health and emotional wellbeing and, in several cases, the easing of symptoms of chronic illness."[56]

Positive cost curve

Lastly, and perhaps most germane, WF had a positive cost curve, producing benefits that far exceeded expenses. The simplest calculation concerns direct benefits from improved efficiency. Presuming the government's numbers can be trusted, and that all households shown by a benefit entitlement check to be eligible for additional benefits capitalized on them, from 2001 to 2011 the WF Scheme cost roughly £2.4 billion, but, with an average annual increase in benefits per household of £1,894.79, it generated £87.2 billion in undiscounted savings for fuel-poor English homes.[57]

However, even this calculation conservatively estimates the true benefits from WF. As part of a national health impact evaluation, the Warm Front

Table 3.5   Cost and years of life saved under England's Warm Front Scheme

Intervention	Cost (£)	Months of life saved	Average cost per life year saved (£) over ten years
Insulation	280	0.26	12,905
Heating	1,130	0.51	26,629
Insulation and heating	1,410	0.56	30,449

Source: Jan Gilbertson

Study Group looked at the costs of the program and the average cost per life year saved. Table 3.5 shows that the combined cost of insulation and upgrading the heating system averaged £1,410. This resulted in an increase in temperature from investments in insulation and heating which added an extra 0.56 months to the lives of a 65-year-old couple living together (0.33 months for the man and 0.22 for the woman). Over ten years the average cost of extending a recipient's life by one year would be £30,449 if heating and insulation is installed. If only insulation is installed, the resulting temperature rise adds an extra 0.26 months to the lives of the couple, but the average cost of extending a recipient's life by one year is much less, at £12,905.[58]

The lesson is that the WF scheme extends lives cost-effectively, a conclusion also confirmed by studies outside England. One survey of energy efficiency measures implemented in Northern Ireland found that investments in reducing fuel poverty had a positive cost curve when one accounted for reduced health care expenditures in houses that had an intervention compared with those that did not.[59] Similarly, a study in New Zealand estimated the value of the health, energy, and environmental benefits of retrofitting insulation into 1,350 low-income buildings and found that total benefits in present value terms were 1.5 to 2 times the cost of fitting the insulation.[60]

Challenges

Despite these benefits, the WF scheme has faced and continues to confront serious challenges. These fall into the categories of (1) rising levels of fuel poverty throughout the country, (2) cost-effectiveness, (3) targeting, and (4) fuel consumption.

Rising levels of fuel poverty

The most apparent challenge is that the WF scheme did not meet the Warm Homes and Energy Conservation Act target of ensuring that by 2010 no vulnerable households would remain in fuel poverty. Instead, national rates of fuel poverty have been increasing, rising from 1.2 million vulnerable homes in 2003 for all of the UK – including England, Scotland, Northern

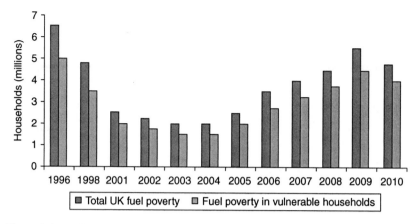

Figure 3.6 Fuel poverty among all UK households and vulnerable households, 1996–2010

Source: Department of Energy and Climate Change.

Ireland, and Wales – to 2.8 million in 2007 and 4.0 million in 2009, when 18 percent of all households in the UK were defined as fuel-poor, as Figure 3.6 illustrates.[61] As one study from the Association for the Conservation of Energy warned, "given the increases in the number of fuel poor households over recent years, the 2010 target has not been met and the 2016 target is in serious jeopardy."[62] Even worse, vulnerable households are still the "hardest hit" by fuel poverty, accounting for 80 percent of fuel-poor households in England.[63] Although households owned by the elderly make up the single largest group living in fuel poverty, almost one-third consist of families with children or those with single parents. Furthermore, rates of fuel poverty in Northern Ireland and Scotland are roughly *double* the rate in England, possibly indicative of the difficulty of eradicating fuel poverty in rural locations, places with inefficient housing stock, and places where energy prices are unusually high.[64]

The causes behind a resurgence of fuel poverty relate to the energy intensity of British homes and rapidly rising energy prices more than any inherent failure with WF. Household energy bills in the UK are dominated by a winter heating season which, due to the country's colder climate, requires 20,000 kWh per year for space heating for a typical three-bedroom domestic house.[65] From 2004 to 2012, Figure 3.7 shows that domestic electricity prices increased by over 75 percent and gas prices rose by more than 122 percent,[66] with gas prices rising 15 percent alone from 2011 to 2012.[67] Every one percent rise in fuel prices results in approximately 40,000 households entering fuel poverty.[68] As the Department for Energy and Climate Change was forced to admit, "For several years, prices have been the most influential factor in movements in fuel poverty. Prices have risen at a rate well above that

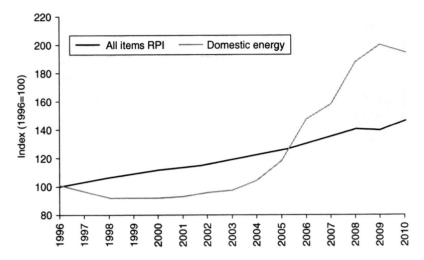

Figure 3.7   Domestic energy prices in the UK and the Retail Prices Index, 1996–2010
Source: Department of Energy and Climate Change.

of income."[69] Unfortunately, analysts expect fuel poverty to worsen in the future. Deutsche Bank has projected that, with expected energy price rises of 25 percent by 2015, a quarter of the country could fall into fuel poverty.[70] This combination of low household incomes, higher fuel prices, and higher penalties for nonpayment will make it exceedingly difficult to contain fuel poverty, let alone eradicate it.[71]

All the while, rather than increase the budget for WF, the government has curtailed it. In a review of government spending in 2010, the government committed itself to a "smaller, more targeted" WF scheme, cut its budget by two-thirds to a total of £210 million for 2011 to 2013, and restricted eligibility to fewer households.[72] The WF scheme will also cease in March 2013 and will be replaced by the Affordable Warmth element of the Energy Company Obligation. The Energy Company Obligation will be funded by energy companies and there will no longer be a national government-funded program to tackle fuel poverty in England. This is unfortunate, to say the least, given that the nonpartisan Centre for Sustainable Energy estimates it will cost at least £4.6 billion, rather than the meager few hundred million pounds currently allocated, to deliver energy efficiency improvements to all poor households that need them.[73] As the *Lancet* recently warned, "the decrease in budget allocated to the Warm Front scheme for 2011–2013 is therefore unlikely to meet demand, even allowing for reduced eligibility."[74] The 2012 Hills Review of fuel poverty estimates that under current trends there will be more than nine million households in fuel poverty, representing almost half (43 percent) of all households in the United Kingdom, by 2016.[75]

Cost-effectiveness

A second class of challenges concerns the cost-effectiveness of WF. One major obstacle was the requirement that homes still pay for a proportion of the energy efficiency work being done when it exceeds the maximum grant level – WF grants do not always cover all of the costs of the work. The average household contribution required in 2007 and 2008, for example, was £581, and roughly one-quarter of applicants that winter were asked to contribute to the cost of upgrades.[76] This meant that more than 6,000 households withdrew from the scheme and a further 16,000 households did not finish their application or put their application on hold for more than a year. Thus, WF hasn't reached everybody it needed to, mitigating its cost-effectiveness.

Furthermore, though the section above on benefits showed how the WF scheme saved households money and energy, it may not be the most cost-effective way to mitigate greenhouse gas emissions. Presuming it saved about 1.5 tons of carbon dioxide per household per year, the WF scheme abated carbon dioxide at a cost of about £50 per ton.[77]

Targeting

The WF scheme had difficulty identifying fuel-poor homes and likely disbursed a significant proportion of its assistance to homes that were not technically in fuel poverty. Tom Sefton from the London School of Economics calculated that in 2002, oddly, only 42 percent of fuel-poor households received assistance under WF and that 75 percent of those participating were not, in actuality, fuel-poor.[78] Two years later Sefton repeated his analysis and found that "just less than one in five Warm Front recipients are fuel poor prior to receiving a grant...most recipients – around four in five – are probably not fuel poor."[79] In other words, WF grants were skewed towards low-income households, but not necessarily fuel-poor households.[80] The recent 2012 Hill Review similarly noted that fuel poverty is difficult to distinguish from "more general problems of poverty."[81]

Targeting difficulties within the WF scheme have been confirmed by two separate evaluations conducted by the National Audit Office. The first estimated that "around a third of the fuel poor may be ineligible and up to two thirds of eligible households may not be fuel poor." It cautioned that the heating and insulation measures under the Scheme would be insufficient to move households out of fuel poverty in at least 20 percent of the cases, and that only 14 percent of WF grants reached the least efficient homes.[82] The second warned that "57 percent of vulnerable households in fuel poverty do not claim the relevant benefits to qualify for the Scheme."[83] Commentators on the internet have joked that, as a result, the scheme should have been renamed the "Lukewarm Front."

Figure 3.8 presents the most likely explanations behind this faulty targeting and shows that the problem is a mismatch between fuel-poor households

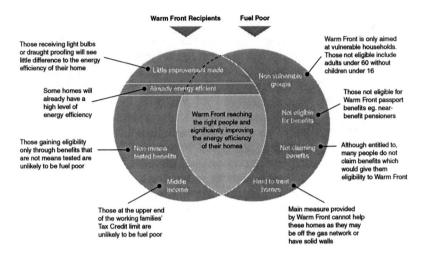

Figure 3.8 Explanations for poor targeting and gaps in coverage under England's Warm Front Scheme

Source: National Audit Office.

and eligibility for WF. WF relies on self-selection for participation, which means many homeowners will not know about the program, or may not consider themselves fuel-poor. For example, surveys have shown that only five percent of homeowners will report that they cannot afford adequate heating when in fact more than 12 percent cannot.[84] The implication is that people don't want to report themselves as fuel-poor, or they believe that they are in fact not fuel-poor. In other cases, people resent the idea of having "strangers" come into their private homes, both to do the initial energy audits and assessments, and later when installing energy-efficient equipment. Also, the elderly are comparatively less adept at discerning temperature, and may feel comfortable at temperatures unhealthy for them.[85] Other low-income households will spend significantly less on fuel than required and suffer cold homes as a consequence. Under the current definition of fuel poverty these households won't show up as fuel-poor because they aren't spending 10 percent of their income on energy services. In other cases, low-income households may already be energy-efficient, meaning they didn't really need assistance under the scheme. In still other cases, they may decline to participate in the WF scheme out of fear of bothering their landlord, or they may want to avoid the transaction costs and inconvenience of having repairs done in their home.

Fuel consumption

A significant number of WF homes, even though they saved their occupants money, did not necessarily consume less fuel. Put another way, a large difference exists between the fuel savings modeled and expected by WF planners

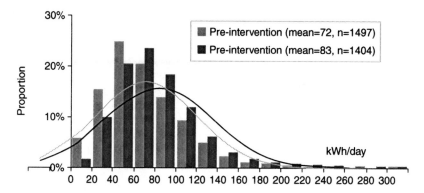

Figure 3.9 Increase in fuel consumption after England's Warm Front measures
Source: Jan Gilbertson.

and actual fuel consumption among participating homes. Research undertaken by the Warm Front Study Group found that WF homes did not always exhibit reduced fuel consumption.[86] Counter to intuition, fuel consumption rose on average after WF measures in many homes studied – findings reflected in Figure 3.9.

One reason for higher fuel consumption in the post-intervention dwellings can be attributed to the "take back" or comfort factor. Households "take back" the benefits of improved energy efficiency as increased warmth and comfort rather than as fuel savings, particularly following the installation of a new heating system.[87] Other reasons for this rising fuel consumption may relate to contractors trying to keep costs low and implementing slipshod work quickly, or to contractors trying to meet the requests of clients to keep costs low and on budget. Another, albeit unconfirmed, factor could be contractors taking advantage of elderly and vulnerable customers unable to distinguish high-quality work from low-quality. The research team, for example, utilized infrared images of 85 WF dwellings and found missing areas of loft insulation in 13 percent of the cases and in cavity walls in one-fifth of the cases. Another reason is higher monitored ventilation rates than predicted, caused by gaps created by retrofitting that allowed heat to escape and because occupants of warmer homes are more likely to open windows to let "fresh" air in, mitigating energy savings.[88] Finally, some residents may experience difficulty in using their new heating systems effectively and may resort to using their older equipment, resulting in less fuel consumption savings than might be expected.

Conclusion: lessons and implications

Even though it faces these thorny challenges and will cease to exist after 2013, the Warm Front's approach towards addressing fuel poverty reveals four noteworthy lessons for other countries and communities.

First, it demonstrates the truly massive financial benefits of investing in energy efficiency. Presuming that the numbers from the Department of Energy and Climate Change are accurate, every £1 invested into WF produced as much as £1 to £36.3 in benefits – and this is potentially conservative, as it excludes indirect savings related to reduced health care expenditures, avoided excess winter deaths, and longer lives for those living in more efficient homes. WF reminds us that energy efficiency programs can pay for themselves quite quickly, and produce measurable benefits that far exceed costs. The Department of Energy and Climate Change reaffirms this conclusion with its most recent carbon cost abatement data, shown in Figure 3.10, which demonstrate that a variety of household energy efficiency measures are the "cheapest" and "best" investments society can make if it wants to mitigate emissions of greenhouse gases.[89]

Nor is this lesson limited only to England and the United Kingdom. One assessment of 41 energy efficiency projects completed in the industrial sector of the United States found that they recovered the cost of implementation in slightly more than two years and then yielded an aggregate $7.4 million in savings every year thereafter. On a global scale, the International Energy Agency reviewed large-scale energy efficiency programs and found that they saved electricity at an average cost of 3.2 ¢/kWh, well below the cost of supplying electricity from *any* source.[90] A comprehensive report from the independent consulting firm McKinsey & Company looked at the maximum potential of an assortment of technological options to abate greenhouse gases (presuming a tax of €60 per ton of CO_2 was in place), and found that

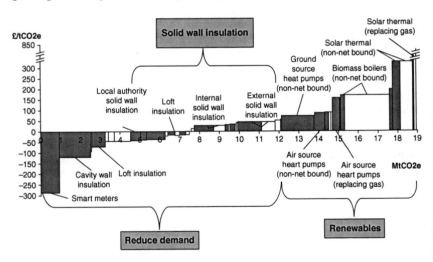

Figure 3.10 Carbon cost abatement curve for various household options in the UK, 2012

Source: Department of Energy and Climate Change.

a host of residential and industrial energy efficiency options were far more cost-effective than building power plants.[91]

Second, the Warm Front program, though it faced some major obstacles, underscores the necessity of having strong political leadership for addressing fuel poverty. Although it did not engender enough political commitment to prevent the recent rise in English fuel poverty, WF was the product of energy planning at the highest levels of the UK government, including the efforts of a 1999 Inter-Ministerial Group on Fuel Poverty, the Warm Homes and Energy Conservation Act in November of 2000, and a 2001 formal government Fuel Poverty Strategy. Managed by the Department for Environment, Food, and Rural Affairs, WF received additional support from a fuel poverty indicator produced through an English House Condition Survey and Census data, and a staggering £2.6 billion in government funding. With this level of support, the WF scheme was able to assist more than two million predominately vulnerable low-income households in improving the efficiency of their homes and creating the economies of scope and scale needed to drive costs down, performing installations more cheaply than other private contractors.

Third, the somber barriers confronting WF remind us how difficult eradicating fuel poverty can be. Though the absolute number of fuel-poor households in the UK would undoubtedly have been higher without WF, the scheme stopped lowering the national rate of fuel poverty in 2004, and since then the percentage of homes in fuel poverty has *tripled* to the point where it applies to roughly one in five throughout England, a number that could rise even further by 2016 to almost *half* of UK households. Some of the culprits behind this rising rate of fuel poverty were outside the scheme's control, such as prices for electricity and natural gas that have rapidly increased compared with incomes. Others, however, can be blamed on the government, such as drastic reductions in the WF budget. And the sensitivity of the number of households in fuel poverty to increases in the price of fuel suggests that energy efficiency might need to come *second* as a priority to government efforts to keep fuel prices low; indeed, WF's largest contributions towards reducing fuel poverty coincided with a fall in the retail price of energy, which dropped to a 20-year low in 2003.

In another way, the current situation in England – rising prices, rising rates of fuel poverty – illustrates the intractability of eradicating fuel poverty without ensuring that household incomes are improved along with their energy efficiency. One implication of this finding is that electricity restructuring and liberalization may be incompatible with attempts to abolish fuel poverty – the first wants people to pay competitive market costs for energy, the second says they shouldn't have to pay full costs if these come to represent a disproportionate share of their income. The replacement to the Warm Front scheme, the Green Deal, is so far suffering from a similar problem. The Green Deal is a commercial low-interest, pay-as-you-save private loan

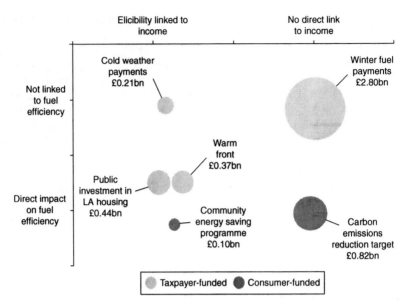

Figure 3.11   Funding for various fuel poverty-related policies in the UK, 2009
Source: Department of Energy and Climate Change.

scheme for domestic energy upgrades, the idea being that consumers can pay back their loans through their fuel bills, with payments always being smaller that the resultant real time savings, free of government expenditures. Yet, as of January 2013, four months after the Green Deal commenced in October 2012, few companies have signed up to offer loans and not a single customer has formally participated in the scheme.[92] In short, "paying" for energy efficiency and relying on market approaches seems incommensurate with meaningfully addressing fuel poverty.

Another related implication of this conclusion is that times of fiscal austerity create competing and contradictory pressures within government; as fuel prices rise, governments tend to have less revenue and curtail public spending, yet cutting back during these times means less support for helping the poor at precisely the time they need it the most. The government also has incentives which counteract addressing fuel poverty and investing in efficiency. Every year, for example, Figure 3.11 shows that £2.8 billion is spent on winter fuel payments to pensioners at Christmas, typically a lump sum of £250 per household – yet only one-quarter of these expenditures were actually spent on energy efficiency improvements.[93] This annual amount is equivalent to the *entire* ten-year budget of WF, and if it had instead been invested into WF the government might have achieved its target of eliminating the scourge of fuel poverty entirely. Some even accuse the UK government of intentionally passing fuel poverty costs onto the consumer to relieve the government's tax system and budget.[94] Yet another implication is that, given the sensitivity of fuel poverty to fuel prices, the government's goal of mitigating climate change (which will cost money and raise prices) could exacerbate the numbers of households in fuel poverty (which need prices to drop rather than rise) if those homes are unable to implement efficiency measures to offset increased prices.

Fourth, and lastly, the WF reminds us how vital human behavior and trust can be in determining whether energy efficiency interventions succeed or fail. Social attitudes played significant roles in mitigating the ability of the WF scheme to meet its targets. The WF scheme was based on self-selection, meaning homeowners themselves had to apply for assistance. Between 42 and 57 percent of vulnerable fuel-poor households did not take advantage of the WF scheme. The largest reason behind this refusal to participate is simple: people either did not consider themselves fuel-poor or, if they were, did not want to admit it. Fuel poverty takes a severe toll on households, and as such it may have become stigmatized to the point where households feel that classifying themselves as "fuel-poor" or even just "poor" is insulting. In other cases, households didn't trust that a government-led program could actually help them, or resisted the idea of having strange auditors and contractors invade the intimacy privacy of their homes. The repercussion is that future measures for tackling fuel poverty must be group-specific, and find ways of convincing families that are frustrated, angry, and scared that they can save energy without sacrificing their social identity, pride, or privacy.

4
Due Process and the World Bank's Inspection Panel

Introduction

Due process – the idea that everyone deserves equal protection under the law, and should be guaranteed a basic list of human rights – is one of the oldest justice concerns, dating all the way back to antiquity and culminating in the United Nations Declaration on Human Rights, which virtually every country signed in December 1948. Yet many large, international energy companies have consistently failed to respect human rights for their workers as well as the communities that they operate in. Indeed, oil and gas suppliers have depended on private security firms to protect their operations and suppress dissent. In Indonesia, Myanmar, Nigeria, and Peru, some firms selling oil and gas have denied free speech, employed torture, supported slavery and forced labor, sanctioned extrajudicial killings, and ordered executions. Shell gave guns to Nigerian security forces, and Chevron provided aid, helicopters, and pilots to an armed group that then gunned down nonviolent protestors on an oil drilling platform. British Petroleum, ExxonMobil, ConocoPhillips, and other companies offer daily "security briefings" for mercenaries and supply vehicles, arms, food, and medicine to soldiers and police throughout the world.[1]

Far from these being isolated examples, this chapter documents a horrifying list of the ways that modern energy production can devastate communities when it occurs without proper oversight. However, it also shows that mechanisms such as Inspection Panels are emerging to hold these actors accountable for their actions. The idea behind such panels is that they make it possible for citizens and communities to "challenge decisions of international bodies through a clear and independently administered accountability and recourse process."[2]

Due process as an energy justice concern

The reality of contemporary energy production is that it frequently abuses human rights, a term referring to basic categories of civil, cultural, economic, political, and social rights and extended to cover issues of property, economic development, human health, safety, and the natural environment.[3]

For example, the oil company Shell paid $15.5 million in settlement after having been accused of collaborating in the execution of eight activists protesting against company activities in Nigeria, including Ken Saro-Wiwa and a grassroots organization called the Movement for the Survival of the Ogoni People. When the group became popular, families of the Ogoni Tribe alleged in their lawsuit that Shell conspired with the Nigerian militia to capture and then hang the men. Documents were also presented indicating that Shell worked with the Nigerian army to bring about the torture of other protestors and that Shell contributed information that resulted in the killing of dissidents. Shell was accused of providing the Nigerian army with vehicles, patrol boats, and ammunition, as well as helping plan raids and military campaigns against villages.[4]

Elsewhere in Africa, Chinese firms have embarked upon an "oil safari," with billions of dollars of oil contracts signed with Angola, Ethiopia, the Congo, Gabon, Kenya, Madagascar, Nigeria, and the Sudan. These firms have followed a "no-questions-asked" policy of "noninterference," meaning they "don't hold meetings about environmental impact assessments, human rights, bad governance and good governance."[5] In Sudan, the Chinese National Petroleum Corporation (CNPC) funneled $4 billion to the regime in Khartoum, enabling them to purchase Shenyang fighter planes and an assortment of heavy arms that they then used to suppress a secessionist movement in the south. The CNPC also permitted the Sudanese government to use its oil platforms to launch attacks, with groups armed with Chinese weapons.

In Asia, the Talisman Oil Company has been directly involved in the civil war in Afghanistan, with evidence of the deployment of mujahedeen and child combatants to provide security around oil and gas blocks.[6] The Unocal Corporation, now Chevron, authorized the use of rape, murder, extermination, forced relocation, and slavery to expedite the construction of oil and gas pipelines in Burma/Myanmar.[7] In both the Karen and Mon states, the Myanmarese government has habitually relied on forced labor to construct pipelines and carried out military attacks on civilians opposing such projects. Unocal admitted in an American court to knowing that the regime in Myanmar had a record of committing human rights abuses, that the military forced villagers to work and entire villages to relocate for the purposes of the project, and that the government committed various acts of violence in front of company employees.

Furthermore, millions of individuals are involuntarily resettled every year due to energy projects. The largest two sources are hydroelectric dams and mining. About four million people are currently displaced by activities relating to the construction of hydroelectricity projects or their operation annually, and 80 million have been displaced in the past 50 years from the construction of 300 large dams.[8] This relocation, however, often proceeds without true and meaningful consultation and consent.

A tragic example comes from Sarawak, the largest state of Malaysia, where plans for the Sarawak Corridor of Renewable Energy (SCORE) would build no fewer than 12 hydroelectric dams connected to industrial facilities along the coast of Borneo. Rather than provide electricity to the local population, the dams will instead supply all of their electricity to factories and smelters. The environmental impact assessments associated with the first of these dams, the 2,400 MW Bakun Hydroelectric Project, were conducted only after construction commenced, and vastly underestimated the facility's negative effect on water quality, ecosystems, and communities. Key decisions to proceed with projects like Bakun and Murum, a 944 MW dam, also occurred without the consent of affected communities, even though Bakun dam necessitated the forceful removal of 10,000 indigenous people and Murum will require the resettlement of 3,400 people. One community leader told the author that:

> In Ulu Bakun, before we were relocated, we needed no money. We could walk out the front of our longhouse and there was the forest, the river, there was everything we needed. Here, we are surrounded, boxed in, blocked. Our way of living has changed forever.

Community leaders don't speak out to authorities, however, because those vocalizing their opposition to SCORE have been reputedly beaten, tortured, and, in some cases, murdered.[9]

Halfway across the world, Turkey, Syria, and Iran have all heavily dammed the headwaters that flow into the Tigris and Euphrates Rivers for electricity and irrigation. The combined result is much less water for Iraq, leaving the Shatt al Arab river without enough supply for livestock, crops, and drinking. Tens of thousands of Iraqi farmers have already had to abandon their fields and homes for lack of water.[10]

In Indonesia, poorly regulated state-owned coal mining enterprises, responsible for most production, have extensively displaced communities without consultation or consent. Part of this involuntary resettlement has to do with Article 33 of Indonesia's constitution, which states that the natural resources of the country are to be exploited under state control – with no need for public input – for the maximum benefit of the people of Indonesia. The Ombilin coalmine, with reserves of 109 million tons, has disregarded local land rights entirely and improperly managed acid rock drainage and sediment ponds to the point where 60 square kilometers of land had to be abandoned by indigenous communities.[11]

Indeed, one wide-ranging international survey estimated that at least 2.6 million people have been displaced due to mining in India from 1950 to 2009, and individual mines in Brazil, Ghana, Indonesia, and South Africa have involuntarily displaced a further 15,000 to 37,000 people from their homes.[12] That study noted two troubling conclusions. First, the impact of such displacement extends beyond loss of land, which represented only 10 to 20 percent of impoverishment consequences, to include joblessness, homelessness, marginalization, food insecurity, increased health risks, social disarticulation, and the loss of civil and human rights. Second, it found that compensation for resettlement rarely occurred, and that, when it did, such compensation was insufficient to restore, let alone improve, quality of life for the displaced.

These types of activities are not always the result of evil oil and gas companies. International donors and banks, such as the World Bank Group, have violated their own guidelines and degraded the rights of communities. In Argentina, local brickmakers relied on the special characteristics of the sediment and sand of the Parana River to create networks of clients and suppliers over hundreds of years, a social fabric that was completely "destroyed" by the construction of the massive Yacyretá Dam. Along the Jamuna River in Bangladesh, the Char people subsisted on shifting sand islands for thousands of years, but had their lifestyles interrupted by the creation of the Jamuna Bridge. In India, thousands of villages in the region of Singrauli have seen their homes "uprooted and resettled" for the establishment of coal mines; those who have protested or resisted have been met with "police brutality and violence." As one study put it:

> What these communities have in common is that their misfortune resulted from development projects funded by the World Bank – projects ironically aimed at benefiting just such poor and disempowered communities... None of these rural communities was informed of, or allowed to participate in, the decisions that would fundamentally change their lives. The underlying Bank projects were typically designed in closed consultations between their country's finance ministries and World Bank economists.[13]

In these examples, none of these people affected by the World Bank's projects were consulted, nor did they receive any compensation for the loss of their lifestyles.

The common element that ties all of these examples together is that they all violated some aspect of due process. Though conceptions of due process date back to antiquity, most historians trace its modern founding to the Magna Carta, a charter signed by King John of England in 1215 declaring that "no free man shall be seized or imprisoned, or stripped of his rights or possessions, or outlawed or exiled, or deprived of his standing in any other way, nor will we proceed with force against him, or send others to do so, except by the lawful judgment of his equals or by the law of the land."[14] Since then such protections have become enshrined in everything from the United States Declaration of Independence in 1776 to the international ratification of the United Nations Declaration of Human Rights in 1948.

Justice scholars have built upon these ideals to build a more elaborate array of protections sometimes known as "procedural justice" or "representative justice."[15] Generally, these protections center on these interrelated justice issues:

- Who gets to decide and set rules and laws, which parties and interests are recognized in decision-making?
- By what process do they make such decisions?
- How impartial or fair are the institutions, instruments, and objectives involved?

Procedural theories of justice are all oriented with process – with the fairness and transparency of decisions, the adequacy of legal protections, and the legitimacy and inclusivity of institutions involved in decision-making.[16]

Such protections are perhaps best encapsulated in the idea of free prior informed consent, or FPIC. FPIC refers to "a consultative process whereby a potentially affected community engages in an open and informed dialogue with individuals or other persons interested in pursuing activities in the area or areas occupied or traditionally used by the affected community."[17] Its main characteristics are that it is "freely given," "obtained before permission is granted," "fully informed," and "consensual."[18] "Freely given" implies that no coercion, intimidation, or manipulation has occurred: that potentially affected people freely offer their consent. "Prior" implies that consent has been sought sufficiently in advance of any meaningful decision to proceed with a project, before things like financing or impact assessments begin. "Fully informed" means that information about the project is provided that covers its nature, size, pace, reversibility, and scope; expected costs and benefits; the locality of areas to be affected; personnel and revenues likely to be involved; and procedures for resolving conflicts, should they occur. "Consent" means "harmonious, voluntary agreement with the measures designed to make the proposed project acceptable."[19] It does not necessarily mean absolute consensus: a majority of 51 percent suffices most of the time, and it does not always mean direct involvement; consensus can occur through representatives, through plebiscites (direct single issue votes), and referenda. It is, however, distinct from consultation – the act of merely discussing a project with a community – because it gives communities the ability to "say no."[20]

Ideally, practicing FPIC means that any type of project that depends on the use of force, requires involuntary action, or contributes to poverty or social instability is unacceptable. It includes the option of communities being able to refuse projects and also ensuring that communities cannot be resettled involuntarily – that resettlement packages are attractive and that communities consent willingly. As the United Nations Commission on Human Rights recently explained it, "free, prior and informed consent recognizes indigenous peoples' inherent and prior rights to their lands and resources and respects their legitimate authority to require that third parties enter into an equal and respectful relationship with them, based on the principle of informed consent."[21] Its underlying principles are that communities receive information and engage in meaningful consultation about any proposed energy initiative and its likely impacts before it commences, and that institutions of energy decision-making are representative, inclusive of indigenous peoples and minorities.[22] FPIC must involve proper representation of communities (including marginalized groups) and true power-sharing.

Case study: The World Bank's Inspection Panel

The World Bank's Inspection Panel meets some of these criteria, and it was established to ensure that the Bank met its own social and environmental safeguards and protected the human rights of the people affected by its projects. For those unfamiliar with the World Bank Group, often called the "World Bank" and abbreviated "WB" for the remainder of this chapter, it is the largest international development bank in the world.[23] The WB is a multilateral institution that provides loans and credit to developing countries to "stimulate social and economic development." Charged with a mandate of alleviating poverty, it provides financial support by itself and through its partnerships with other institutions to projects that can have significant social, economic, and environmental consequences.[24] Though it operates independently, the WB's major shareholders are France, Germany, Japan, the United States, and the United Kingdom, and its major borrowers are Brazil, China, India, Indonesia, Mexico, and Russia. The WB's annual average lending volume for fiscal year 2012 was $52.6 billion in loans, grants, equity investments, and loan guarantees.[25]

The WB holds a unique place in the world of international relations, as its membership includes almost every country in the world.[26] All of its capital stock is owned by its member states, and the size of a country's share in the Bank is relative to the size of that country's economy. The world's largest economies, known as the Group of Eight, control almost half of the shares in the Bank, the United States being the largest contributor with roughly 17 percent of its shares. Member countries give the WB capital that is then invested in international financial markets to maximize returns; the revenue from these investments is channeled into loans to governments or corporations for projects intended to raise standards of living.

The WB is actually comprised of five separate organizations. The International Bank for Reconstruction and Development (IBRD) was created in 1944 at the Bretton Woods Conference, initially as a special agency of the United Nations. Its purpose was to allocate money from wealthy nations to those that needed help financing reconstruction efforts after World War II. Now, its role has shifted. According to the IBRD website:

> The IBRD works with its members to achieve equitable and sustainable economic growth in their national economies and to find solutions to pressing regional and global problems in economic development and in other important areas, such as environmental sustainability. It pursues its overriding goal – to overcome poverty and improve standards of living – primarily by providing loans, risk management products, and expertise on development-related disciplines and by coordinating responses to regional and global challenges.[27]

The IBRD remains the Bank's largest lender, with $20.6 billion invested in 93 operations in fiscal year 2012 involving 188 member countries, offering money and guarantees to middle-income governments for development projects. Generally, IBRD loans carry an interest rate of 6.98 percent with a maturity of 15 to 20 years as well as a grace period of five years.

The second largest part of the Bank is the International Development Association, or IDA, established in 1960 to fund projects in developing countries unable to borrow money on IBRD's terms. The IDA offers interest-free loans known as "credit" with a maturity of 35 to 40 years, a ten-year grace period and a 0.75 percent service charge. Bankers sometimes call these loans "soft" since they have extended grace periods and minimal finance charges. As the WB describes it, the IDA "supports countries' efforts to boost economic growth, reduce poverty, and improve the living conditions of the poor." In 2012, 81 countries received IDA assistance worth roughly $15 billion in total.

The third branch of the WB is the International Finance Corporation (IFC), created in 1956 and owned by 176 member countries. In 2011, it invested $18.7 billion across 100 countries. Unlike the previous two branches, the IFC lends to the private sector rather than governments, and its mandate is to "promote productive and profitable private enterprises in developing nations" and to allow "financial institutions in emerging markets to create jobs, generate tax revenues, improve corporate governance and environmental performance, and contribute to their local communities." The IFC generally offers technical assistance to companies, including privatizing government-linked monopolies, protecting securities, and creating stock exchanges.

The fourth and fifth arms of the Bank are the Multilateral Investment Guarantee Agency (MIGA), created in 1988 with about $1.1 billion of investments and guarantees in 2011, and the International Centre for the Settlement of Investment Disputes (ICSID), created in 1966 to resolve disputes between foreign investors and governments.

Due to its size and scale, the WB is a major financier of energy projects around the world. As Figure 4.1 shows, energy and transport are among its

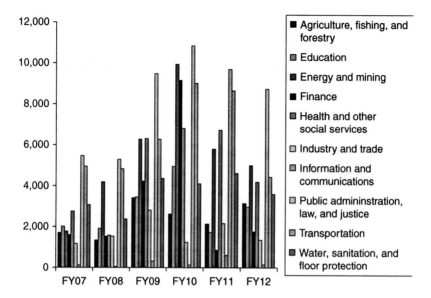

Figure 4.1 World Bank lending by theme and sector for the IBRD and IDA, 2007 to 2012 (millions of dollars)

Source: World Bank Group.

Table 4.1 World Bank Group energy portfolio by financing source, 2007–2012 (millions of dollars)[29]

Institution	2007	2008	2009	2010	2011	2012
World Bank	2,137	4,778	6,648	10,367	6,064	5,634
IBRD	552	2,427	3,569	8,140	4,755	3,017
IDA	1173	1,932	2,155	1,356	1,200	1,853
Climate Finance	287	333	363	619	100	301
Others	125	86	561	253	9	463
IFC	1,308	2,782	1,650	2,354	1,998	2,054
MIGA	417	110	33	225	119	489
Total Energy Financing	3,862	7,670	8,332	12,947	8,181	8,177

Source: World Bank Group.

largest sectors, along with public administration and water and sanitation. Tables 4.1–4.3 show its lending patterns for the past five years.[28] During this period, its energy lending has jumped considerably from $3.8 billion to $8.2 billion, Europe and Central Asia, Sub-Saharan Africa, and South Asia accounting for the bulk of its financing. It has also started lending more for energy efficiency and renewable energy projects, though it does still support fossil fuels and upstream oil, gas, and coal production.

Table 4.2   World Bank Group energy portfolio by region, 2007–2012 (millions of dollars)

Region	2007	2008	2009	2010	2011	2012
Sub-Saharan Africa	1,224	1,261	1,754	5,281	1,156	1,813
East Asia and the Pacific	251	1,505	1,229	990	2,116	968
Europe and Central Asia	518	1,194	2,295	1,182	2,384	2,331
Latin America and the Caribbean	489	1,157	801	1,948	1,331	551
Middle East and North Africa	368	360	806	1,050	67	748
South Asia	947	2,158	1,446	2,495	1,062	1,710
Multiregion Projects	65	35	–	–	64	55
Total Energy Financing	**3,862**	**7,670**	**8,332**	**12,947**	**8,181**	**8,177**

Source: World Bank Group.

Table 4.3   World Bank Group energy portfolio by sector, 2007–2012 (millions of dollars)[30]

Sector	2007	2008	2009	2010	2011	2012
Energy Efficiency	753	1,521	1,685	1,802	1,551	1,353
Renewable Energy	840	1,471	1,678	1,905	2,977	3,615
Fossil Fuels	364	1,087	987	4,287	290	690
Policy Support and Technical Assistance	717	1,015	1,702	2,019	1,783	1,369
Transmission and Distribution	458	1,650	1,204	2,208	1,397	270
Upstream Oil, Gas, Coal	729	972	1,076	725	182	880
Total Energy Financing	**3,862**	**7,670**	**8,332**	**12,947**	**8,181**	**8,177**
Total Low Carbon	**1,761**	**3,338**	**3,363**	**5,584**	**5,937**	**5,480**
Total Access	**905**	**1,784**	**2,201**	**1,020**	**1,031**	**1,499**

Source: World Bank Group.

A final way the WB influences national and international energy decisions, apart from its financing, is through its circulation of technical assistance, capacity-building efforts, and analysis. This has convinced some
scholars to go so far as to call it a "global development agency" and "knowledge bank" for its $25 million "research" budget dedicated to the creation
of information and data presented in books, papers, and journals.[31] In line
with its role as a knowledge bank, the WB also exerts enormous weight
on macroeconomic policy advice and has used its influence to restructure
entire economies more in line with its prescriptions for proper economic
governance.

The role of the World Bank Inspection Panel, hereafter the "IP," is to
operate "as an independent forum to provide accountability and recourse for
communities affected by Bank-financed projects, to address harms resulting
from policy noncompliance, and to help improve development effectiveness of the Bank's operations."[32] The IP decides whether the Bank complies
with its own policies and procedures, especially its social and environmental
impact assessments, and ensures its projects avoid harm to people and
the natural environment. In the words of James D. Wolfensohn, a former
President of the WB:

> When the Board of Directors of the World Bank created the [IP, it was]
> an unprecedented means for increasing the transparency and account-
> ability of the Bank's operations. This was a first of its kind for an inter-
> national organization – the creation of an independent mechanism to
> respond to claims by those whom we are most intent on helping that
> they have been adversely affected by the projects we finance. By giving
> private citizens – and especially the poor – a new means of access to
> the Bank, it has empowered and given voice to those we most need to
> hear. At the same time, it has served the Bank itself through ensuring
> that we really are fulfilling our mandate of improving conditions for
> the world's poorest people. The Inspection Panel tells us whether we
> are following our own policies and procedures, which are intended to
> protect the interests of those affected by our projects as well as the
> environment.[33]

To accomplish these tasks, the IP operates on four key principles: independent fact-finding, problem-solving, checks and balances, and transparency. To respond to complaints from communities affected by WB projects,
the Panel conducts independent investigations and compiles facts to estab-
lish whether the WB followed operational policies and procedures. It plays
a role in helping affected communities document problems with projects
and then considers solutions to them. The IP provides a check and balance
on the WB Board and the actions of Bank managers. It lastly promotes
transparency in Bank operations through the publication of reports and
findings.

History

The Executive Board established the IP in September 1993 as an "independent accountability mechanism" to commence operations on August 1, 1994. The stated goal of the IP was to provide a forum "for people who believe that they may be adversely affected by Bank-financed operations to bring their concerns to the highest decision-making levels of the World Bank."[34] The germination of the IP began in the 1980s when nongovernmental organizations began criticizing the WB for violating its own policies concerning involuntary resettlement (established in 1982), tribal peoples (created in 1982), and environmental assessment (formed in 1988).

The Sardar Sarovar Dam and Canal projects on the Narmada River in India served as a flashpoint for these concerns, as they involved the forced relocation and resettlement of more than 120,000 people and induced significant environmental damage. Initial project plans, among other things, involved the construction of 30 large dams, 135 medium-scale dams, and 3,000 smaller dams on the Narmada River. The largest of these, the Sardar Sarovar Dam, submerged 37,000 hectares of land and required the evacuation and resettlement of 245 villages, almost all of them home to indigenous peoples, across the states of Gujarat, Maharashtra, and Madhya Pradesh. The Canal network, similarly, necessitated the clearing of 80,000 hectares of land for construction.[35] The World Bank approved funding for these projects throughout the 1980s but continued to make disbursements even after civil society groups raised social and environmental concerns. By the early 1990s, the WB was "under attack" by activists and governments for violating its safeguards for these projects,[36] making it what one study called "a lightning rod for transnational protest."[37]

As a response to the ensuing outcry, the WB President Barber Conable created an independent commission to review the Indian projects in 1991 to be headed by Branford Morse, a retired senior administrator from the United Nations Development Program, and Thomas Berger, a former Supreme Court Justice from British Columbia, Canada. The commission released a report, informally known as the "Morse Commission Report," in 1992 and identified "serious compliance failures" by the WB as well as "devastating human and environmental consequences of those violations."[38] Follow-up investigations across the WB's entire portfolio implied that managers habitually failed to carry out the organization's goals of poverty alleviation and environmental protection.[39] Even the WB's own 1992 internal review of its lending practices, known as the Wapenhans Report, concluded that the WB was "suffering from a performance crisis" with almost 40 percent of projects scoring "unsatisfactory" ratings, widespread defaults on loans, and an overall "culture of approval" that prioritized making loans at the expense of local communities and preserving ecosystems.[40] Part of the explanation was that managers were rewarded for moving forward as many projects as possible but were not penalized if those projects suffered from poor design or shortcomings in accommodating local peoples. The implication was that the WB's violations of its procedures symbolized a systematic failure on the part of Bank management.[41]

To assuage its member countries and borrowing governments, the WB created its IP to consist of three separate "circles." The first, and central, circle consists of the three Members of the IP, each from a different country, chosen for a period of five years that cannot be renewed. Members are selected and appointed for their "ability to deal thoroughly and fairly with the requests brought to them, their integrity and their independence from Bank Management."[42] The second circle consists of an Executive Secretariat and a small number of support staff created to help advise and assist the Members in executing their duties. The third circle consists of internationally recognized experts who assist the IP with specific investigations, providing Members with reliable information in fields as diverse as anthropology, forestry, and hydrology.

The IP itself was given three explicit investigatory, advisory, and rule-making powers. The IP has the authority to investigate complaints about the failure of WB projects as they relate to its operational policies and procedures in the design, appraisal, implementation, or evaluation of a project. It has the authority to advise the Executive Directors of the WB about which complaints to pursue. It has the authority to determine which rules apply to a particular case and to resolve any issues that may arise as complaints go forward.

In carrying out these tasks, each complaint brought before the IP passes through four phases: eligibility, investigation, report writing, and board discussion. Any two or more individuals affected by any WB project can bring a complaint before the panel by writing a short letter, formally called a "Request for Inspection." These requests need not be long – some are only one page – they can be in any language, they can be hand-written, they can be extensively or sparsely detailed, and they can be submitted confidentially. Once a Request for Inspection is submitted, the Panel must determine whether it meets the following criteria for eligibility, last updated in 1999:

- The affected party must consist of any two or more persons with common interests or concerns and who are in the borrower's territory;
- The request must assert in substance that a serious violation by the Bank of its operational policies and procedures has or is likely to have a material adverse effect on the requester;
- The request must assert that its subject matter has been brought to Management's attention and that, in the requester's view, Management has failed to respond adequately demonstrating that it has followed or is taking steps to follow the Bank's policies and procedures;

- The matter cannot be related to procurement;
- The related loan must not be closed or substantially disbursed;
- The Panel must not have previously made a finding on the subject matter or, if it has, the request must state that there is new evidence or circumstances not known at the time of the prior request.

Complaints meeting these criteria generally enter a field investigation phase in which the IP retains outside experts and collects facts relating to the case, including visits to the project and meetings with the requesters – the author

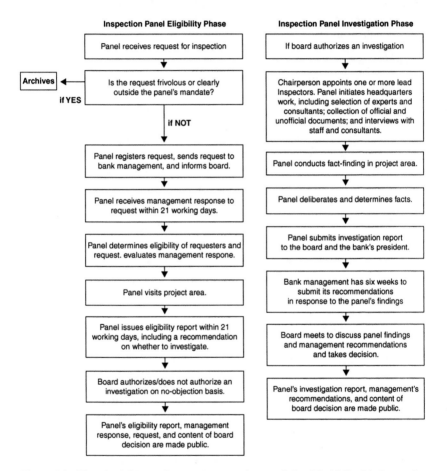

Figure 4.2   The eligibility and investigation phases of the World Bank's Inspection Panel

Source: World Bank Inspection Panel.

says "generally" because in rare circumstances the Executive Board has failed to approve IP-recommended investigations. All data collected during the investigation are synthesized into a report submitted to the Bank's Board, made available to the public upon their approval. The final phase consists of the Board asking relevant project managers to provide recommendations in response to the Panel's report and, when necessary, creating an "Action Plan" to remedy harm and bring the project into compliance. Figure 4.2 graphically depicts the IP process during the first two phases.

Benefits

Between 1994, when it started operations, and June 2011, its most recent evaluation, the IP had received 73 requests for inspection that resulted in 27 approved investigations. Figure 4.3 illustrates that these have mostly dealt with IBRD and IDA loans, with the most common complaints involving supervision and environmental impact assessments.[43] In 2011, the IP dealt with 14 cases, five resulting in investigations, the other nine resolved without complaints. To date, four of the IP's benefits appear most significant: it has (1) improved the governance of WB operations, (2) resulted in the cancelation of harmful WB projects, (3) empowered communities with the knowledge they need to seek meaningful redress and compensation, and (4) enhanced the protection of human rights by starting a hopeful trend in international law.

Improved efficacy of World Bank operations

In theory, the IP has great potential to improve internal WB governance by spotlighting missing information, uncovering flaws in assessments, acknowledging noncompliance, and enlightening management. As previously mentioned, it does this by expanding check-and-balance mechanisms so that concerns can be brought directly before the Executive Board of the Bank, and by creating a public record of how well the institution is complying with its own safeguards.[44] The IP generates high-quality, timely, objective information about projects that can be used by Bank managers wishing to mitigate problems, as well as member countries concerned about the legitimacy of the Bank and governments considering whether they want WB assistance. The public record created by IP investigations and findings can also enable members of civil society to better assist marginalized groups and learn about Bank practices and procedures.[45]

In practice, examples abound confirming that the IP has meaningfully improved the performance of WB managers and the efficacy of programs. In its investigation of the West Africa Gas Pipeline Project, for example, the IP documented that the WB did not meet its own requirements for livelihood restoration and resettlement, and that there had been an embarrassing factor of ten mistake in undercompensating indigenous peoples for the value of their land.

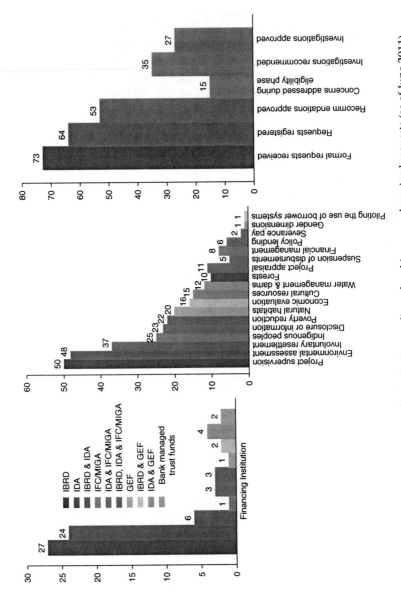

Figure 4.3 Profile of Inspection Panel requests by institution, policy-related issues, and received requests (as of June 2011)

Source: World Bank Inspection Panel.

It also found that project documents overestimated the amount of natural gas flaring the project could reduce – mistakes that were promptly corrected.[46] In its investigation of the Chad-Cameroon oil pipeline, the IP spotlighted how the project did not incorporate a proper regional health plan into its assessments, failed to develop sound water and sanitation facilities around the project area, and lacked an appropriate weighing of how oil and gas projects impacted local communities – mistakes that were also corrected.[47]

In other cases, investigations led to the sanctioning or firing of WB staff. In its investigation of the Lake Victoria Environmental Management Project in Kenya, the IP discovered that project managers had failed to meet the public consultation components of the Bank's environmental impact assessment policy, claims that, when substantiated, led to the formal and public rebuke of bank staff.[48] In its investigation of the Albania Integrated Coastal Zone Management and Clean-Up Project, the IP discovered that managers had intentionally misrepresented facts concerning the demolition of homes in project appraisal documents and board meetings, findings that convinced the Board to "clean house" and terminate the employment of some senior managers.[49]

Suspension and cancellation of harmful projects

IP investigations have successfully suspended and cancelled socially and environmentally harmful WB projects. Strikingly, the best example here is the first investigation ever conducted, of the Arun III Proposed Hydroelectric Project in Nepal in 1994. The IP's investigation of the $800 million project revealed that the Bank had failed to observe proper requirements for relocation and resettlement, and that the project had tenuous economic justification given the fragile state of Nepal's economy. After receiving the IP's report, Bank President James Wolfensohn completely terminated the Bank's support.[50] Analogously, after the IP revealed that the Bank had "extensively violated" its policies concerning project design and the participation of vulnerable ethnic groups in consultations related to the China Western Poverty Reduction Project in Qinghai, the Board cancelled the Bank's involvement. More than a decade later, in 2005, the IP's report had a similar effect on the Mumbai Urban Transport Project in India, a scheme involving the relocation of more than 120,000 people. Its report documented "significant compliance failures in resettlement activities," ultimately convincing the Bank to suspend disbursements until employment options were expanded for local shopkeepers, and until databases for redressing grievances were improved.[51]

The severity of cancelling or suspending projects is credited with creating a prophylactic effect on Bank management. The threat of an IP investigation has contributed to "allergic reactions" among Bank staff to risky or controversial projects, offering an antidote to the "culture of approval" so widespread throughout the Bank in the 1990s. One senior managing director of the WB has stated that, due to the IP, "risk aversion is widespread, among front line managers especially ... It comes out in the choice of projects. It looks like staff are avoiding certain types of projects."[52] Law professor Daniel D. Bradlow has also found that "the possibility that Bank staff could be held personally accountable for the manner in which they make and implement decisions" has encouraged them to be "more responsive to and respectful of those who are affected by their decisions," essentially deterring bad behavior.[53] Moreover, the IFC and MIGA have adopted environmental and social policies and created the Compliance Advisor and Ombudsman Office in 1998 and 1999, respectively, to proactively ensure that their projects better meet WB guidelines.

Broader community empowerment

The IP has empowered communities by giving affected persons stronger voices, respecting their account of events, and redressing implementation problems. As Law Professor Dana L. Clark explains:

> The indirect effects of the Panel process can be very valuable for local citizens affected by Bank projects, as several past cases have demonstrated. First, the Panel process provides a forum where they can raise concerns that have in many cases been ignored for years. Regardless of the ultimate outcome of the Panel process, this raises awareness of the problems at the highest levels in the Bank and in the borrowing country. It can also be empowering for the claimants. Another indirect effect is increased media attention and support from international NGOs interested in Bank activities. International attention has so far played a critical role in pressuring the Board to take action. In addition, the time and effort involved in launching an Inspection Panel claim has increased solidarity among the claimants, empowered them to have a dialogue with government officials and project authorities, increased awareness within the country, and strengthened the networks of support at the local, national and international levels.[54]

Indeed, one independent assessment evaluated 28 claims submitted to the IP between 1994 and 2002 and found that in ten of those cases – Arun, Planafloro, Jamuna, Yacyretá, Jute Sector, Itaparica, Singrauli, Land Reform, China/Tibet, and Structural Adjustment Argentina – the claim had positive impacts on local communities. It concluded that "the Inspection Panel has given increased legitimacy to the claims of local people affected by the World Bank, and it serves as a forum through which their voices have been amplified within the institution."[55]

In Africa, the IP's investigation of the Democratic Republic of Congo (DRC) Forest-Related Operations project drew media attention to the plight of the Pygmy peoples and led to recognition of the noneconomic value they placed on Congolese forests. That recognition has convinced the federal government to place nationwide restrictions on logging.[56] In India, IP investigations of coal mining in Singrauli resulted in increased compensation for affected people; in the case of the $1.7 billion 3,100 MW Yacyretá dam, IP findings resulted in changes in design for smaller reservoirs that preserved the hunting and fishing grounds of some local communities.[57] In Bangladesh, the IP prompted controversy over the Jamuna Bridge, resulting in the national government recognizing, for the first time, the right of the Char people to compensation.[58] The mere act of filing a complaint in the Romania Mine Closure and Social Mitigation Project was enough to motivate the government to resolve it, without the IP needing to conduct an investigation.[59] IP actions have also resulted in legal protections for the Yorongar people in Chad, the virtual elimination of prostitution along the Chad-Cameroon oil pipeline, and greater protections against deforestation in Cambodia.[60]

Enhanced protection of human rights

The IP has motivated other agencies and institutions to create their own accountability mechanisms and enhanced the global protection of human rights. Law Professor Daniel D. Bradlow opines that "the creation of the Panel is an important legal development" since it is "the first forum in which private actors can hold an international organization directly accountable for the consequences of its failure." This prompted the Inter-American Development Bank to create its own inspection panel in 1994 and the Asian Development Bank to create its own inspection mechanisms in 1995.[61] The European Union established the International Right to Know Day on September 28, 2000, and created its own accountability mechanisms, similar to the IP, known as the European Ombudsman Office (EOO). Each year, the EOO receives a sobering 2,500 to 3,000 complaints addressed to institutions of the EU, of which 600 or so are investigated.[62] Similarly, the European Investment Bank established its "Complaints Mechanisms Principles" in 2010 to hold that bank more accountable to its stakeholders; the EIB's mechanism investigates roughly 40 cases every year.[63] As such, Professor of Law Laurence Boisson de Chazournes argues that "the measures taken by the Bank to strengthen information disclosure have contributed to the promotion of transparency and access to information," and that as a result "the Inspection Panel is a progressive step in the development of both the law of international organizations and the law of human rights."[64] Confirming these points, an independent evaluation from the Center for International Environmental Law, Bank Information Center, and the International Accountability Project, hardly fans of the WB, acknowledged that the IP process was useful for protecting human rights and that the Panel's cases "opened the door" for groups to incorporate human rights concerns into their complaints.[65]

Challenges

Though the World Bank's IP has achieved these benefits, it also faces challenges related to (1) transaction costs, (2) a limited mandate, (3) lack of independence and resistance from the WB, (4) retaliation against complainants, and (5) the overall momentum of Bank culture.

Transaction costs of bringing a complaint

Though the criteria for filing a complaint with the IP are relatively simple, and complaints can theoretically be only one page long, in practice filing a successful complaint is a technical process that requires the assistance of experts, including lawyers and development specialists. As one study noted, this process "requires a fair amount of technical knowledge and work on the part of the claimants...the requirements have made access to the panel difficult for those very people it was established to serve."[66] As one example, the only formal point of contact after complainants have contacted the IP is to meet with the Panel if it decides to conduct a field investigation, though, if that does happen, the Panel nearly always visits the country to meet with the requesters, verify signatures, and get their input before writing its report as to whether to recommend an investigation. The requesters can also provide information to the Panel throughout the complaint process, yet there is no opportunity for complainants to comment on how the Executive Board or Bank managers respond, nor are claimants given access to information before decisions are made about their claim, nor do they have a right to comment on whatever remedial measures are to be taken. There is also no right for claimants to appeal Panel findings or the Board's decision about how to proceed with claims and Action Plans.[67]

The end result is that claimants can go through the arduous process of documenting and filing a complaint, only to have nothing done, even if they have demonstrated harm. For instance, in a claim concerning the Lesotho Highlands Water Project, a scheme involving at the time the construction of Africa's largest dam, the project substantially increased water prices for local people and disproportionately impacted poor townships – but the IP decided not to pursue an investigation since "the claimants had not made a link between the conditions they complained of and specific bank policy violations."[68] Similarly, in a complaint filed concerning the $167 million Rondonia Agricultural, Livestock, and Forestry (Planafloro) project, a collection of 25 organizations representing indigenous peoples, small farmers, unions, and environmental groups exhaustively documented how the scheme compromised the integrity of local rainforests. The IP found that the Bank "failed to implement the project as planned" and had, by its poor oversight, facilitated illegal logging which was destroying up to 400,000 hectares of rainforest each year. The Executive Board, however, denied an investigation.[69] (In the IP's defense, this particular case came before the 1999 revisions which changed the Board approval procedure for an investigation to approval on a nonobjection basis, rather than requiring affirmative Board approval.)

Limited mandate

Though the IP can bring attention to problematic projects and has the authority to report findings to the Bank's Executive Board, it has an extremely limited mandate. Most significantly, the IP does not have the ability to provide any relief to complainants, even successful ones deemed worthy of remediation. The IP panel cannot make recommendations, either; it can only make "findings," which may or may not include substantive and important observations. The IP cannot offer compensation, it cannot by itself issue an injunction against further work on a project, it does not oversee the generation of Action Plans, and it cannot rule that a project should be cancelled. It cannot even prevent governments or other institutions from retaliating against complainants, something discussed in greater detail below. The IP merely advises the Executive Board, which then determines how to proceed.[70] As Lori Udall, Senior Advisor to the Bank Information Center, explains, "the Panel is – by the nature of its powers – a limited creature. It is not an enforcement or judicial mechanism."[71]

This inability to provide relief has proven to be a substantial barrier. In the case of the 3,100 MW Yacyreta Hydroelectric Project in Argentina and Paraguay, although the IP found that the Bank "had violated numerous policies and procedures," Bank management never fully followed through to ensure the complete implementation of Action Plans, though management did report back to the Board on at least two occasions.[72] In the case of the Cartagena Water Supply, Sewerage, and Environmental Management Project in Colombia, the IP found "numerous problems" with how the project would discharge waste into the Caribbean Sea. Yet it took Bank management three years to implement an Action Plan – by which point most of the damage to coastal villages had already been done.[73] And, in the case of the Mumbai Urban Transport Project, Bank management failed to meet the deadline for submitting an Action Plan and delayed for more than a year, meaning almost none of the targets recommended by the Executive Board had been met.[74]

Moreover, the IP is constrained by the criterion that that it cannot investigate projects where "the loan financing the project has been substantially disbursed." This is problematic for two reasons. First, many problems with Bank projects do not manifest themselves until years after disbursement. Second, bank managers who know or suspect a complaint is coming can strategically disburse large amounts in order to make it "nonredressable," manipulating the process to their advantage.[75]

Even when IP investigations do attract media attention, their impact can be short-lived. As one review recently noted:

> The short-term benefits that come from the added attention brought by filing a Panel claim do not necessarily translate into long-term sustainable benefits. Too often, the Panel's recommendations and the subsequent Board decision provide momentum for change – momentum that is lost once the Panel's and civil society's attention turns.[76]

Lack of independence and internal resistance

The independence of the IP, and its acceptance within both the Executive Board and the Bank as a whole, has ebbed and flowed. At least two distinct phases are discernible: a period of fierce resistance after the IP's first case involving the Arun Dam in Nepal, and a period of almost universal support. The first six years of the IP were certainly tumultuous, and witnessed intense and protracted resistance to the IP within the Executive Board. Between the Arun Dam investigation in 1994 and 1995, and the China Western Poverty Reduction Project in 1999 and 2000, five years lapsed and extensive reforms were made before the Executive Board would yet again authorize the IP to conduct full investigations. During those five years, in every case where the IP recommended an investigation, the Board, upon advice from a combination of senior managers, borrowing country governments, and some donors, either rejected the IP's authority to investigate (in cases such as Planafloro, Itaparica, and EcoDevelopment) or limited the terms of reference for the investigation (in Yacyretá and Singrauli). During this phase, Law Professor Dana Clark notes that:

> Bank management resented the panel's scrutiny and strongly resisted being held accountable. Management responses to the Inspection Panel claims tended to deny policy violations and deny responsibility for problems identified in the claim (usually blaming the borrowing government for the problems instead). Management responses also tended to challenge the eligibility of claimants and propose "action plans" as alternatives to panel investigations.[77]

Throughout this rough period of adjustment, Kristine J. Dunkerton, Executive Director of the Community Law Center in Maryland, called the IP nothing more than a "public relations tool for the Bank and a political liability for borrowing nations."[78] Admittedly, the Executive Board of the Bank was in a precarious position: they had to balance their attempts at impartiality and support for the nascent IP with the wishes of managers and donors with whom it had to work.[79]

Things started to change in 1998 when the IP entered a second phase of almost universal acceptance. Most Board members were "frustrated that a mechanism that had been designed to help reduce their problems in dealing with difficult projects was instead generating more conflict at the Board."[80] Consequently, in March 1998 the Board began a formal review of the Panel process, establishing a working group consisting of three borrowing country executive directors and three donor country executive directors.[81] The review's conclusions, released in April 1999, called on the Board to make a number of reforms, which it did. The Board released an internal memorandum to all staff affirming the "importance," "independence," and "integrity" of the IP. It instructed project managers and staff to base their recommendations on the findings of the Panel, rather than to try and preempt Panel investigations. Most meaningfully, it reached a compromise concerning the issue of the Board rejecting IP findings for investigation by clarifying that "if the Panel so recommends, the Board will authorize an investigation without making a judgment on the merits of the claimants' request, and without discussion," except with respect to technical criteria for eligibility. The Panel's recommendation on whether to conduct an investigation was also sent to the Executive Board on a nonobjection basis. In exchange, the IP's assessment period was reduced to simplify the evaluation process.[82] Testing these changes, the complaint over the China Western Poverty Reduction Project showcased the Board fully respecting the IP's recommendation for a full investigation, and, for the next decade or so, the board "universally authorized" the panel to investigate when it so recommended.[83] During the 2010 and 2011 fiscal year, even though four of the IP's reports on eligibility required full discussion by the Board instead of their usual approval on a "nonobjection" basis, the Board has respected the IP's autonomy.[84]

Retaliation against complainants

Perhaps the most serious challenge faced by the IP is retaliation against those who file complaints, in some cases involving imprisonment and even death. In the case of the Mumbai Urban Transport Project, for example, one of the lead requesters was imprisoned shortly after the IP sent its report to the Board (though he was released a few months later). In the case of the Chad-Cameroon pipeline, security forces from the government repeatedly threatened requesters in an attempt to persuade them to drop the complaint.[85]

In the case of the coal mines in the Singrauli region of central India, retaliation was more severe. The WB had loaned $150 million to the National Thermal Power Corporation (NPTC) in India to enable it to build the Singrauli Super Thermal Power Plant, but environmental and social assessments found that 90 percent of those having to be relocated had already been displaced once. An independent report warned that the previous resettlement "appears to have failed in practically all cases" and highlighted "the inadequacy of facilities and equipment necessary for water supply, sewage treatment, schools and education, and medical care."[86] In 1993, still more loans came into India, with $400 million in financing for the expansion and upgrading of the Rihand and Vindhyachal coal-fired power plants, making the NPTC the single largest borrower in the history of the WB at that time. Facing threats of abuse and eviction, a group of families banded together and filed a complaint with the IP in 1997. The act of filing the claim, however, prompted a "retaliatory backlash from NTPC, which moved into the project area with police force and heavy machinery," and "in the area around the Vindhyachal power plant, people were beaten and physically restrained while their homes were bulldozed in the presence of and at the behest of NTPC."[87]

Bank culture

Though the IP has improved the efficiency, efficacy, and accountability of WB operations, it has not addressed some of the deeper underlying problems at the institution. Countervailing pressures, such as incentives that reward staff for lending large amounts of money and meeting annual lending targets, remain.[88] There is evidence that most managers remain focused on "dollars lent" rather than "poverty reduced," and staff are still incentivized to move projects through the approval process faster, both to raise revenue and to avoid offending borrowing governments.[89]

There is also evidence that the WB is a culturally and intellectually homogeneous organization that may be incapable of understanding, let alone protecting, indigenous people of different cultures. For instance, eight out of every ten employees at the WB are trained in economics and finance institutions in the United States and United Kingdom, rather than in sociology or anthropology, or at leading institutions in the developing world. As one study critiqued, "this homogeneity in staffing poses some difficulties for institutions trying to open themselves up to higher levels of participation, involvement and ownership by those most affected by their programs."[90]

Sociologist Michael Goldman has gone so far as to attack the WB's overly technocratic approach to development, arguing that, for every $1 invested by the WB in developing countries, somewhere between $10 and $13 goes back to its Members. A greater amount of capital therefore flows *out* of borrowing countries rather than *in*. And, since the WB always has more capital than it can lend, to survive Goldman argues that it must continually grow its investment portfolio and create demand for its services. Put in this light, the WB's primary task is not to be held accountable for its projects, or to help poor people, but to make money as a business.[91] Others have attacked the WB for contributing to the erosion of local cultures around the world.[92]

As an advisory body, the IP cannot address these fundamental concerns. As one study criticized:

> As an interior body of the Bank itself, [the IP's] ideas cannot be completely independent of the ideology of that institution. Because the Panel is an arm of the Bank, it is by definition an institution with a de facto World Bank bias and consequently acts with the interests of the institution in mind and not necessarily with the interests of the affected communities.[93]

Put another way, even if the IP makes the WB more accountable, it does not significantly reform the Bank's central mandate to finance loans and return investments to its Member Countries.

Conclusion: lessons and implications

At least three conclusions emerge from the WB's Inspection Panel. The first is that improving transparency – broadly defined as a "process by which information about existing conditions, decisions, and actions is made accessible, visible, and understandable" – has immense value in enhancing the internal governance of an institution and its ability to achieve social and environmental goals.[94] The IP has seen WB managers self-monitor their own behavior; it has seen peers, other donors, and governments monitor WB actions to ensure it is meeting its standards; it has seen communities and civil society groups monitor WB actions on the ground. It has reduced errors and at times corruption within Bank management, and made it harder to conduct shoddy analysis or rush projects to meet deadlines and lending targets. Transparency from the IP has facilitated the involvement of displaced and indigenous communities in WB policymaking through formal Action Plans, and functionally made the WB more accountable to the lives of those its projects impact.

The second, however, is that, for transparency to truly occur, the institution promoting it must be independent. One needs strong, committed leaders on all sides. In the case of the IP, numerous presidents of the Bank and members of the Board had the foresight to create an independent accountability mechanism. Members of the IP had the fortitude to stand up to the Board and senior management when they resisted such accountability. Communities and their leaders exhibited the strength and dedication to file complaints, even in the face of potential retaliation and, at times, violence. Such communities benefitted from remarkably brave social activists.[95] Put another way, transparency can be a wonderful mechanism but only if matched with institutions and champions devoted to it succeeding.

Third, the IP affirms the weight of due process and procedural justice as an energy justice concern. Procedural concerns such as the fairness and transparency of energy decisions, the adequacy of legal protections, and the legitimacy and inclusivity of institutions involved in energy decision-making can matter as much as the substantive concerns raised in other chapters of this book. Due process intersects with recognition (who is recognized), participation (who gets to participate), and power (how power is distributed in decision-making forums). Though procedural justice may strike some as dry and unimportant, the IP reminds us that fair procedures matter because they tend to promote better – more equitable, but also more efficient and effective – outcomes.

5
Information and the Extractive Industries Transparency Initiative

Introduction

Roughly 3.5 billion people live in countries with plentiful oil, gas, and mineral reserves, yet a worrying number of these countries do not release transparent information about the extraction of those resources or how the revenues emanating from them are expended. Furthermore, the challenge of governing energy resources is that countries with large caches of them can perform worse in terms of economic growth, social development, and good governance than other countries with fewer resources. The mismanagement of their resources, sometimes called the "resource curse," expands outside the extractive industries sector and slowly degrades social welfare. Resource-cursed countries consequently display comparatively higher levels of poverty and inequality, deteriorating environmental quality, institutionalized corruption, and an increased frequency of conflict and war. After introducing readers to these concerns, this chapter shows how transparency – "timely and reliable economic, social and political information accessible to all relevant stakeholders"[1] – can counteract the trend towards the resource curse. It then presents a case study of the Extractive Industries Transparency Initiative to demonstrate how access to information about energy revenues can improve the accountability of governments, the reputation of corporations, national investment climates, the empowerment of communities, and the promotion of transparency as a global norm.

Information as an energy justice concern

Emil Salim, the President of the Extractive Industries Review, an attempt by the World Bank to improve transparency and accountability in the mining and energy sectors, once began a conference on governance with a curious statement: "Not only have the oil, gas, and mining industries not helped the poorest people in developing countries, they have often made them worse off."[2] Juan Pablo Perez Alfonzo, one of the founders of OPEC, was even pithier in his take on the value of oil to the national economy, calling it "the devil's excrement."[3] Both statements imply that for many countries with relatively rich endowments of energy resources, ranging from coal and minerals to gas and crude oil, those very commodities can become "curses" to the government and its citizens.

Indeed, a long line of political economists and theorists have stated different variants of the "resource curse" argument, which explains that natural resource endowments can incite, prolong, and intensify government failure and violent conflict.[4] Chatham House scholar Paul Stevens argues that large windfall revenues from oil or gas production can change government behavior, damage economic growth, and imperil development.[5] Stanford political economist Terry Lynn Karl postulates that the consolidation and extraction of natural resources can create and solidify asymmetries in wealth that then contribute to rising income gaps between the rich and poor, institutionalize corruption, and enable oppressive regimes to maintain their political power.[6] Political scientist Indra De Soysa notes that civil strife and social instability are strongly associated with natural resource abundance, particularly with mineral, gas, and oil exports.[7] Political geographer Marilyn Silberfein states that resource-based conflicts can become some of the most intractable because participants often benefit from unstable conditions, which can facilitate access to resources, smuggling, and covert trade. Beneficiaries not only grow in their power but also become reluctant to terminate bloody conflagrations.[8] Canadian geographer Philippe Le Billon suggests that resources can be used to finance conflicts (such as the civil war in Sierra Leone over diamonds), or can be a motive for other countries to go to war (such as the Iraqi invasion of Kuwait in 1990).[9]

Californian political scientist Michael L. Ross elaborates three separate strands of the resource curse theorem. The "cognitive" strand implies that windfalls from resource extraction produce myopic disorders among policymakers. The "social" strand suggests that windfalls empower recipients who then favor growth-stunting and self-serving policies. The "statist" strand argues that windfalls weaken the state institutions necessary for long-term economic development.[10] Many investigations weave all three strands together. The inverse of the resource curse also appears justifiable: lack of natural resources has not prevented many countries such as Japan, Hong Kong, Singapore, South Korea, and Taiwan from achieving booming export industries based on solid manufacturing sectors and rapid economic growth.[11]

The problem, in essence, is that many of the most valuable resources, such as diamonds, gold, oil, and gas, are worth so much money that their extraction can be severely mismanaged but still create net profits for companies and governments.[12] Resource-cursed governments tend to manage natural resources opportunistically rather than strategically. That is, government officials view them as a mechanism to create immediate wealth rather than a long-term but depletable asset that should be carefully managed over time. Resource depletion, therefore, tends not to serve the public interest, but instead consolidates wealth among a very narrow class of bureaucrats and managers. Moreover, the immense value of those resources can mask mismanagement and inequity. As one high-ranking executive for a large transnational oil company told the author, "high prices [for natural resources] cover a multitude of sins."

The easiest resources to control – the ones most germane to the resource curse – are "point" resources, resources spatially concentrated in small areas, relatively easy to monitor, and extracted through capital-intensive technologies. "Point" resources contrast with "diffuse" resources, which are spatially distributed and therefore more "lootable", exploited by less capital-intensive measures, and more difficult to consolidate.[13] Point and diffuse resources can contribute to different patterns of capital ownership and distribution of resource benefits. While both can be corrupted, or rapidly extracted or pillaged to serve limited ends or means, point resources are more prone to the resource curse.

Because of their spatially concentrated nature, point resources such as oil and gas are well attuned to highly centralized control. Ruling elites can extract them by controlling small geographic areas, and the revenues generated by even modest levels of extraction can be significant.[14] Diffuse resources, such as coal, charcoal, wood, diamonds, gems, cobalt, gold, silver, sapphires, rubies, and other small, strategic, or valuable metals, are more difficult to track and govern due to their distributed nature. While sapphires and rubies provided the Khmer Rouge in Cambodia and the Karen in Burma with significant revenues during the 1990s, their extraction required almost complete authoritarian control of vast amounts of land.

Five distinct and interrelated conditions appear to explain the structural social, political, and economic factors behind the resource curse.[15] First, those countries most susceptible to the resource curse have economies that are not diversified and remain heavily dependent on a few extractive industries – some of them, such as Iran and Nigeria, receive more than 80 percent of their national revenue exclusively from oil and natural gas.[16] When the price of these few commodities fluctuates widely, the sudden price gyrations and boom and bust cycles erode budgetary discipline, decrease expenditures for public sector investment, and complicate efforts at state planning. Figure 5.1 reveals that oil prices, for example, have been incredibly volatile from 1968 to 2012. Between 1972 and 1992, regions with high primary mineral and oil exports experienced periods of trade volatility two to three times greater than average industrialized economies over the same period.[17]

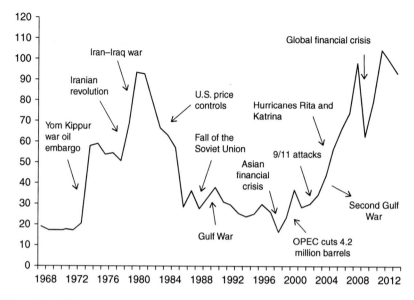

Figure 5.1   Global crude oil prices, 1968–2012

Source: Adapted from US Energy Information Administration. Note that prices are in constant 2011 US dollars per barrel.

Standard practices of banks and financial institutions can serve to exacerbate the volatility of natural resource commodities. When commodity prices fall or interest rates rise, lenders are quick to call in their loans (a general maxim of bankers is that they always prefer to lend to those that do not need their money). When prices fall, resource-dependent countries also need more money, but this is usually when banks want their money back. Capital flows and the decline of oil prices, for example, tend to be pro-cyclical, worsening the fluctuations brought about by the initial fall in the price of oil. Falling prices also tend to incentivize increased amounts of resource production to maintain stable levels of revenue, precipitating greater environmental damage, economic concentration, and depletion of the resource. When prices are high, leaders face expectations that they should increase spending, embark on capital-intensive projects, and reward productive components of the economy with subsidies – further entrenching them.[18]

Second, mineral and commodity exports can cause economic consolidation and the "Dutch Disease," or rapid inflation and increases in the exchange rate of local currency, rendering other exports less competitive.[19] Economists and historians call this the "Dutch Disease" because North Sea oil and gas production had rapid negative effects on the exchange rate and inflation in the Netherlands in the 1960s. In essence, the primary resource being exported then crowds out other sectors of the economy, causing even more economic concentration. In response, regulators often adopt policies that protect that very sector to a greater degree, reinforcing dependence on the resource and worsening the cycle. The risk of economic consolidation becomes greater the more valuable the natural resource, for, unlike other sources of wealth and manufactured goods, lucrative minerals need only be extracted rather than produced to attain revenue. This creates "enclaved" spheres of economic activity without major linkages to other sectors and can take place without large-scale participation of the labor force.[20]

Nigeria offers an excellent example of this "crowding out" in practice. The discovery of oil there is relatively new, occurring in 1956. Since then, however, Nigeria has completely transitioned from an agrarian, diversified economy to one dependent on petroleum products. From 1960 to 2012, oil output rose from a mere 5,100 barrels per day to 2.7 million barrels per day, and government revenue from oil rose from 66 million naira ($417,000) to more than 10 billion naira ($61.3 million) per day. Currently, oil accounts for 40 percent of Nigeria's GDP, 95 percent of its exports, and 83 percent of its government revenue.[21]

Third, most extractive industries rely on sophisticated technology not owned or licensed by the host countries. Oil and gas production and mining, for instance, are very capital- and technology-intensive, meaning they create fewer jobs per dollar invested and require unique skills not well suited for other areas of economic activity. Resource-rich countries often send highly skilled labor abroad to train, and foreign workers are typically sent in to do the local extracting, minimizing the chance that exporting countries will "learn by doing" or own the intellectual property and technology associated with extraction.

Fourth, since most extractive industries and technologies are large-scale, capital-intensive and foreign-owned, few productive links exist between resource extraction and the rest of the economy. Generally, the revenues flow directly to the government in the form of royalties, taxes, and profits. This consolidates revenue among the existing political parties, worsens the gap between rich and poor, and also breaks the link between taxation, representation, and accountability, since most resource-rich regimes do not need to tax their own people (or are less obsessive about collective taxes). The consolidation of revenue also increases the chances of regulatory capture and rent-seeking, since only a small number of officials need to be "captured" to facilitate large transfers of wealth. The concentration of wealth from resource revenues sometimes enables rulers to spend their money on subsidies to friends, family, and political supporters. Terry Lynn Karl, for example, estimates that from 1974 to 2000 OPEC states spent 65 to 75 percent of their income in this manner.[22] (For more on how this environment is prone to corruption, readers are invited to read Chapter Six).

Fifth and finally, large extractive projects frequently fail to benefit minorities and poor communities because these stakeholders have not participated in project selection and financing (for more on this trend, see Chapter Four). The decision-making processes in resource-cursed countries tend to lack transparency and accountability, so vested interests determine what gets built and who benefits. Local communities and indigenous populations must then contend with the damages from resource extraction while the political and corporate elite reap the benefits. Moreover, when those in power conduct cost–benefit analyses, benefits are usually overestimated publicly while costs are underestimated. In the absence of effective oversight and regulation, companies lower their social and environmental standards to reduce costs and maximize short-term returns. In the hazardous extractive industries sector, the consequence of poor oversight can be severe, including acute and catastrophic spills, explosions, accidents, and disasters along with chronic and gradually deleterious exposure to noxious emissions, noise, degraded habitat, and food and water contamination.[23]

For instance, in May, 2006 exploratory natural gas drilling in East Java, Indonesia, ignited mud laced with methane and induced a volcano that grew into an "apocalyptic, flammable sea" that engulfed dozens of villages.[24] Some days enough mud erupted to fill 50 Olympic Size swimming pools, nearby sewers caught fire, and unwitting residents tossing their cigarette butts ignited seeping methane, leading to destructive explosions. In addition, more than 50,000 people lost their homes, and $400 million in damages occurred.

Perhaps the extraction of oil and gas resources in Angola, Columbia, Nigeria, Saudi Arabia, and OPEC as a whole best exemplifies how the resource curse can occur at multiple scales of governance, in many degrees, and in a collection of different countries.

In Angola, which has extracted some of the world's largest oil and gas reserves for 40 years, average life expectancy has plunged to 36.8 years in 2006 (due to a combination of poverty, high rates of diseases such as malaria and cholera, and food insecurity), the infant mortality rate has risen to 19 percent, the country imports most of its food, and almost three-quarters of the population live below the poverty line.[25] Proceeds from oil and gas production have nourished oppressive tendencies within the state. Oil revenues have enabled elites to capture virtually all economic enterprises, and encouraged violent contests for political power from a dissatisfied population. The United Nations has accused Angola of using state-used oil revenues to fund blood diamond operations.[26]

In Columbia, oil and gas profits have contributed to acts of terrorism and intensified military conflict. Guerrilla groups reap an oil "war tax" and earn $140 million per year from oil-related extortion and kidnappings. The National Liberation Army (ELN) and Revolutionary Armed Forces of Columbia (FARC), the two largest rebel groups, have extracted war taxes from oil companies and local contractors through threatened sabotage and murders, extortions, and bombings. When their threats have not gotten them the leverage they desire, ELN and FARC guerrillas have dynamited state oil and gas pipelines more than a thousand times in 13 years, spilling 2.9 billion barrels of crude oil and greatly damaging ecosystems and local water resources. The economic losses alone from explosions on the Cano Limon-Covenas pipeline amounted to $1 billion from 1990 to 1995, or seven percent of Columbia's entire export revenue over the same period. People have joked that the pipeline has so many holes they call it "the flute." Paramilitary groups have also built a cottage industry by stealing natural gas and gasoline by drilling holes in pipelines. Ecopetrol, one state oil company, estimates they lose $5 million per month due to these sorts of losses. Finally, oil and gas revenues provide the Columbian Army with significant income needed to purchase weapons and train soldiers. A military tax of roughly $1 per barrel (or $12 to $30 million per year) allows the army to increase its troop presence and prolong its struggle against the rebels, meaning that oil funds both sides of the conflict.[27]

In Nigeria, the 13th largest producer of petroleum in the world, oil production has generated billions of dollars in government revenues, but the country's average GDP in the past decade was less than its GDP in the 1960s and the poverty rate increased from 27 percent in 1980 to 70 percent in 1999.[28] Per capita incomes actually declined in constant dollars from $250 in 1965 to a paltry $212 in 2004. Between 1970 and 2000, the number of Nigerians living on less than $1 per day grew from 36 percent to more than 70 percent, or from 19 million to 90 million. Rather than bringing peace and stability, oil production has exacerbated political flashpoints. A collection of smaller Niger Delta states have attempted to expand their access and control over oil and gas resources, struggling with minority groups clamoring for sovereignty. Militant youth movements have grown from the resulting insecurity and promote intensified campaigns of interethnic violence.[29] Development economist Uwafiokun Idemudia has therefore concluded that Nigeria is witnessing "the full manifestation of the resource curse."[30] From 1960 to 1999, Idemudia argues that Nigeria lost $380 billion to "corruption and mismanagement" of the oil sector, and that in 2003 70 percent of its oil wealth was "either stolen or squandered."

Saudi Arabia is heavily dependent on oil production, and derives 85 percent of its total export revenue from the oil industry, so much so that every dollar increase in the price of a barrel of oil means a gain of about $3 billion to the Saudi treasury. The Saudis have not used their oil wealth to improve social welfare, however, but have instead consolidated it among a very small number of families. Five extended families in the Middle East own about 60 percent of the world's proven crude oil reserves, and the House of Saud controls more than a third of that amount. Per capita income in Saudi

Arabia has fallen significantly from 1981 to 2001, and seven out of ten jobs in the country are filled by foreigners (and close to nine out of every ten private sector jobs). The disparity in income between the rich and poor has generated significant resentment towards the royal family, a resentment sharpened by belief that the Saudis have failed to protect Muslims in Palestine and elsewhere and have been corrupted by their money. Adding to the social instability, Saudi Arabia has one of the highest birth rates in the world (37.25 births for every 1,000 citizens, compared with 14.5 in the United States), half the population is under 18, and half is unemployed.[31]

Finally, Terry Lynn Karl analyzed the impact of oil production on the political economies of Venezuela, Nigeria, Mexico, Iran, Algeria, and Norway and found that in only one, Norway, did oil extraction correlate with positive development. In each of the other oil-exporting countries, bottlenecks in production, declines in other economic activity (especially in the agricultural and industrial sectors), capital flight, double-digit hyperinflation, and sudden declines in living standards occurred. Karl hypothesizes that these countries have become "addicted" to oil, creating lavish subsidies for oil companies and pursuing oil profits over the public good. She concludes that production and dependence on oil has changed the entire political economy of each country. The per capita GDP for all OPEC members from 1965 to 1998, for instance, decreased by an average of 1.3 percent each year, whereas per capita GDP for all other developing countries that did not rely on oil exports as a whole grew by an average of 2.2 percent over the same period.[32]

Under the right conditions, fortunately, information and transparency can partially counteract some aspects of the resource curse. New Zealand philosopher Gillian Brock argues that the ability to be informed about what is happening to one's environment, the opportunity to have meaningful input into social affairs and decisions that affect one's community or wellbeing, and the ability to seek justice through an independent judiciary when such liberties are violated are an instrumental part of preserving basic fundamental liberties.[33] Principle 10 of the 1992 Rio Declaration on Environment and Development, proclaimed by the United Nations at the same time as the United Nations Framework Convention on Climate Change was being signed, stated the global right to fair and transparent information as follows:

> At the national level, each individual shall have appropriate access to information concerning the environment that is held by public authorities, including information on hazardous materials and activities in their communities, and the opportunity to participate in decision-making processes. States shall facilitate and encourage public awareness and participation by making information widely available.[34]

Furthermore, the Aarhus Convention, which entered into force in October 2001 and has since been signed by 40 European and Central Asian countries, clearly stipulates the necessity of information. Article 7 states that:

> Each Party shall make appropriate practical and/or other provisions for the public to participate during the preparation of plans and programs relating to the environment, within a transparent and fair framework, having provided the necessary information to the public.[35]

Similar statutes and claims supporting public access to information about energy and the environment appear in the 1992 Convention on Biological Diversity, the 1993 North American Agreement on Environmental Cooperation, the 1994 United Nations Convention to Combat Desertification, and a series of bilateral treaties.

Access to information and transparent frameworks for preserving that access have been known under certain conditions to reduce corruption and improve social stability. One study looked at the correlation between corruption, defined as the abuse of public office for private gain, and transparency, defined as access to information, in 150 countries.[36] As Figure 5.2 shows, it found a "certain" correlation between lack of transparency and high levels of corruption. It noted, among other findings, that:

- a lack of transparency makes corruption less risky and more attractive;

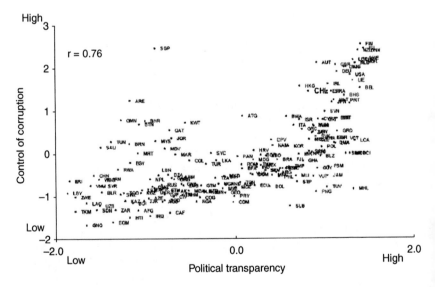

Figure 5.2   Political transparency and corruption for 150 countries, 2009
Source: Kolstad and Wiig.

- a lack of transparency makes it harder to use incentives to make public officials act honorably;
- a lack of transparency makes it hard to select the most honest and efficient people for public sector positions or as contract partners;
- a lack of transparency makes cooperation more difficult to sustain, and opportunistic rent-seeking more likely.

As just one example proving their point, in Uganda only 13 percent of grants for education actually reached schools throughout the 1990s, the rest being "captured" by local government. As soon as the Ugandan government began publishing the details of those grants in newspapers, 80 percent reached those schools.[37] A similar, second study of 105 countries from 1960 to 2004 confirmed that "lack of transparency" can significantly and negatively impact economic growth. That study documented such a strong "causal relationship" across various sample sizes and timeframes that it concluded that "the lack of transparency...is one of the primary reasons for the subsequent poor growth record for these countries."[38]

Case study: The Extractive Industries Transparency Initiative

The Extractive Industries Transparency Initiative (EITI) offers one relatively effective approach towards promoting revenue transparency for the oil, gas, and mining sectors. Put simply:

> The EITI is a multi-stakeholder initiative involving multinational and state-owned extractive companies, host governments, home governments, business and industry associations, international financial institutions, investors and civil society groups, which have established a broad consensus on the ways and means of revenue transparency. The EITI emphasizes the prudent use of natural resources wealth and dictates that the management of such wealth should be exercised in the interests of national development.[39]

The EITI operates on the principle of having free, full, independent, and active assessments of the ways that extractive industries companies interact with government and impact communities and society.[40]

History

The roots of the EITI extend back to at least 1999, when the nongovernmental organization (NGO) Global Witness released an influential report called *Crude Awakening* detailing the negative impacts of oil and gas production in Angola.[41] The report persuaded Transparency International, another NGO, to form an alliance with Global Witness and numerous other NGOs known as the "Publish What You Pay" (PWYP) campaign. That campaign called for action not only in Angola but in other major oil and gas developers to ensure that the revenues from these sectors were more transparent. However, many oil companies saw PWYP as too demanding, and it ran into significant opposition from governments. To push forward transparency in the oil sector, in 2002 the billionaire philanthropist George Soros put his weight behind the campaign and in September of that year, at the Sustainable Development summit in Johannesburg, South Africa, then British Prime Minister Tony Blair announced that he would be creating a less stringent, voluntary scheme to "promote the transparency of oil, gas, and mining revenues." [42] This scheme came to be known as the EITI in June 2003, when a high-level meeting in the United Kingdom consisting of representatives of governments, industries, and civil society groups endorsed Blair's approach and agreed upon a common set of "EITI Principles." These principles were later endorsed at the annual summit of the Group of Eight, an international consortium of the world's largest economies, in 2004. [43]

One year later, in 2005, a formal set of "EITI Criteria" was agreed upon with further support from the World Bank and International Monetary Fund, and an International Advisory Group was appointed with the task of managing the EITI process. [44] An interim International EITI Secretariat was formed within the UK Department for International Development (DFID), and in January 2005 it developed a Sourcebook of "Guidance for EITI Implementation." The International Advisory Group, chaired by Transparency International founder Peter Eigen, met five times throughout 2005 and 2006, quickly published an "EITI Validation Guide," and by the end of 2006 the EITI had a multi-stakeholder board and the support of a permanent secretariat to "manage the EITI at the international level." [45] Near the end of 2006, the EITI was registered as a formal not-for-profit organization in Norway, and a new and expanded "EITI Association" was adopted at the Doha Conference in February 2009. [46]

The key to the EITI approach is its "multi-stakeholder" take on transparency, involving three distinct sectors – government, civil society groups, and corporations in the extractive industries. [47] In its most up-to-date form, the EITI sets six fundamental stipulations. First, it demands the "regular publication" of "all material" oil, gas, and mining payments by companies to governments ("payments") and all material revenues received by governments from oil, gas, and mining companies ("revenues"). This publication must be disseminated to a wide audience in a publicly accessible, comprehensive, and comprehensible manner. Second, when such audits are lacking, payments and revenues are to be subject to a "credible, independent audit" of reputable "international standards." Third, reporting of payments and revenues is to be reconciled by an "independent administrator" which identifies and corrects discrepancies. Fourth, no companies are to be exempt from EITI reporting, meaning it covers private companies, public state-owned companies, and hybrid government-linked companies. Fifth, the active engagement of civil society is required in the design, monitoring, and evaluation of the EITI process. Sixth, the public is to be kept informed by the timely publication of "work plans" for how the host governments will manage their revenues, implement EITI reforms, and assess capacity constraints.[48]

To meet the "EITI standard," countries must fulfill a sobering 21 separate requirements, shown in Figure 5.3. To become an EITI "Candidate," a country must meet five "sign-up" requirements. It then must publish an "EITI Report" within a year and a half that reconciles what companies say they pay in taxes and royalties with what governments say they have received. To achieve EITI "Compliant" status, a country must complete an "EITI Validation" which provides an "independent assessment of the progress achieved and what measures are needed to strengthen the EITI process." If a country has made "meaningful" progress but does not yet meet all of the requirements for Compliance, the EITI Board can decide to renew its "Candidate status,"

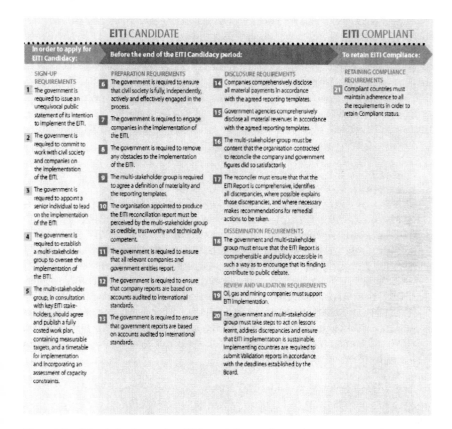

Figure 5.3   Criteria for becoming EITI candidate and compliant countries, 2012

Source: Adapted from EITI Secretariat.

but a country may hold Candidate status for no more than 3.5 years. When no progress is "meaningfully made," the EITI Board can revoke a country's Candidate status.[49] The EITI process has a degree of feedback and flexibility built into it. An implementing country can petition the EITI board to review any decisions about its "Candidate" and "Compliant" status.[50]

Benefits

As of early 2013, 18 countries were "EITI Compliant": Azerbaijan, Central African Republic, Ghana, Iraq, the Kyrgyz Republic, Liberia, Mali, Mauritania, Mongolia, Mozambique, Niger, Nigeria, Norway, Peru, Tanzania, Timor-Leste, Yemen, and Zambia. Nineteen other countries were "EITI Candidates": Afghanistan, Albania, Burkina Faso, Cameroon, Chad, Côte d'Ivoire, the Democratic Republic of Congo, Gabon, Guatemala, Guinea, Indonesia, Kazakhstan, Madagascar, Republic of the Congo, São Tomé and Príncipe, Sierra Leone, Solomon Islands, Togo, and Trinidad and Tobago. In total, some 29 countries have disclosed payments and revenues in an "EITI Report," and 60 of the largest oil, gas, and mining corporations in the world actively participated in the EITI process. The EITI has also received support from 84 global investment institutions that collectively manage about $16 trillion in energy infrastructural assets.[51]

The relatively broad coverage and participation in the EITI compared with other voluntary schemes relates to the perceived benefits it brings to governments, companies, and communities. As University of Illinois College of Law scholar A. Friedman put it:

> Compliance and candidacy under the EITI has a vast array of benefits to both countries and corporations. First, compliant and candidate countries use their membership to strengthen the investment climate. It is a signal to investors and financial institutions that there will be increased transparency, accountability and governance. It is also possible that this promise will reduce violent conflict around the natural resource sectors. For corporations and investors, doing business in EITI Compliant countries reduces both political and reputational risk. This, in turn, reduces costs by reducing the need for or lessening the cost of risk insurance. As for the general population, it is generally advantageous to have more information in the public arena through transparency, as well as the benefits associated with greater foreign direct investment.[52]

Indeed, the following six benefits appear to be the most significant: (1) access to information, (2) governmental accountability, (3) corporate responsibility, (4) improved investment climates, (5) community dialogue and empowerment, and (6) the maintenance of transparency as an emerging global norm.

Access to reliable information

The simplest benefit, also the bedrock for all of the EITI's other advantages, is the creation of reliable and timely information about oil, gas, and mining in a given country. As of January 2013, the number of EITI Reports has jumped from zero to almost 150 in seven years. Figure 5.4 shows how, as of 2012, 30 countries have disclosed data from 109 fiscal periods with the participation of more than 900 companies. Some countries, such as Nigeria, have disclosed information dating back to the 1990s; others, such as Azerbaijan, continually update their data every year. All reports cover oil and gas, but some

Country	Status	Fiscal periods disclosed	Reporting companies	Published in	Covering years	Covering sectors
				LAST REPORT PUBLISHED		
Albania	Candidate	1	34	2011	2009	● ●
Azerbaijan	Compliant	8	31	2011	2010	● ●
Burkina Faso	Candidate	2	4	2011	2008-2009	●
Cameroon	Candidate	8	19	2010	2006-2008	● ●
Central African Republic	Compliant	4	9	2010	2007-2009	●
Congo	Candidate	7	16	2011	2010	●
Côte d'Ivoire	Candidate	2	5	2010	2006-2007	●
DR Congo	Candidate	1	26	2010	2007	● ●
Gabon	Candidate	3	30	2008	2006	● ●
Ghana	Compliant	6	10	2011	2009	●
Guinea	Candidate	1	6	2007	2005	●
Iraq	Candidate	1	37	2011	2009	●
Kazakhstan	Candidate	5	123	2011	2009	● ●
Kyrgyz Republic	Compliant	6	26	2011	2009	●
Liberia	Compliant	3	71	2011	Jul 2009-Jun 2010	● ● ✹
Madagascar	Candidate [2]	3	3	2010	2007-10	●
Mali	Compliant	4	40	2011	2009	● ●
Mauritania	Candidate	5	36	2011	2009	● ●
Mongolia	Compliant	4	101	2011	2009	● ●
Mozambique	Candidate	1	6	2011	2008	● ●
Niger	Compliant	5	68	2011	2007 -2009	● ●
Nigeria	Compliant	10	27	2011	2006-2008	● ●
Norway	Compliant	2	72	2010	2009	● ●
Peru	Candidate	7	51	2012 1	2008-2010	● ●
Sierra Leone	Candidate	2	9	2010	2006-2007	●
Tanzania	Candidate	1	11	2011	Jul 2008-Jun 2009	●
Timor-Leste	Compliant	2	17	2011	2009	● ●
Yemen	Compliant [2]	3	14	2010	2006-08	● ●
Zambia	Candidate	1	16	2011	2008	●
Total		**109**	**917**			

Figure 5.4   Summary of EITI country reports, 2006–2012
Source: Adapted from EITI Secretariat.

extend to mining and even forestry and agriculture, such as Liberia's 2009 Report, or the Central African Republic, which includes small-scale artisanal mining. Ghana, Mongolia, and Peru have gone beyond the minimal requirements of reporting revenue and have also published data on all financial flows to local governments.[53]

Governmental accountability

The EITI has benefitted governments by enabling them to follow "an internationally recognized transparency standard" that signifies a commitment to "reform" and "anti-corruption." As one independent assessment surmised, "the presence of a legal and institutional framework on transparency can be used by citizens and civil society as a valuable tool to hold their governments accountable on how revenue is spent."[54] The publication of EITI reports, for example, has strengthened public knowledge about extractive industries and established a foundation for improvements. In Nigeria, the EITI Secretariat notes that:

> Nigerian Extractive Industries Transparency Initiative (NEITI) reports, which are publicly available, contributed to significantly better transparency in Nigeria's oil industry, collecting and publishing an array of detailed and useful information for the first time. Nothing remotely like this has been done before, let alone published. The reports went far beyond the basic core requirements of global EITI; NEITI produced not only raw data on the industry and on tax and other fiscal matters; but it also provided crucial and useful insights into processes involved in the industry that have helped many insiders and outsiders to see the oil sector in overview for the first time.[55]

There, the National Assembly passed into law the Nigeria Extractive Industries Transparency Initiative Act in 2007, which made Nigeria the first country in the world to legally require the public reporting of all extractive industries revenues. Before the EITI, for some years it was common for as much as $800 million in oil and gas taxes, royalties, and signature bonuses to go "missing," amounts equal to the individual budgets for the Ministries of Education, Health, and Power.[56] The EITI has since been credited with auditing government expenditures to the point where "more than $1 billion [has been channeled] into government resources that otherwise would have been siphoned off due to corruption."[57] As Professor Humphrey Asobie, Chairperson of the Nigerian EITI's National Stakeholders Working Group, explained, "by placing embarrassing facts and figures about the bulk of Nigeria's public revenue in the public domain, NEITI has become both an instigator of civic interrogation of public officers and a safety valve – redirecting youthful energies from resorting to violent conflicts to engagement in civil debate on sensitive issues."[58]

Similarly, Ghana's EITI Report documented that the agencies responsible for collecting mining payments did not collaborate, and that disbursement records were suspect, leading to better accounting records from the government. In Azerbaijan, discrepancies in the inflows to the government were discovered to be an error in rounding and calculation, mistakes that were quickly corrected.[59] When Azerbaijan began its EITI process in 2003, audits revealed double-digit discrepancies between corporate receipts and government intakes (implying theft and misallocation), yet in 2009 the difference became "nonexistent."[60] Liberia's first EITI Report in 2003 documented a missing $30 million in taxes, royalties, and land rentals, yet their most recent report noted discrepancies in receipts of less than $8,000.[61]

Corporate responsibility

The EITI benefits companies by creating a "level playing field" and enabling them to better engage with community leaders and civil society groups. All too often, Transparency International founder Peter Eigen notes, "companies believe they must adopt unfair or even illegal practices because they feel threatened by competing operators vying for contracts in countries that have unstable or unreliable institutions regulating the sector."[62] The EITI stops this "race to the bottom" by making good governance and transparency the norm rather than corruption, bribery, and rapaciousness.

Furthermore, EITI participation can augment corporate reputation, enabling companies to "recruit the best young people" who are increasingly concerned about social issues, and enabling potential investors to support a company with "positive corporate characteristics."[63] Law professor David Hess notes that the EITI empowers shareholders to act as "surrogate regulators," and that:

> Disclosure enables stakeholders to hold the corporation accountable by comparing the corporation's stated goals to its actual performance and the performance of other corporations. It also helps to improve other corporations' performance because disclosure allows stakeholder groups, including other similarly situated corporations, to examine solutions to the same problem.[64]

Moreover, greater transparency can lessen political risk for investors, since more open business dealings create less opportunity for accusations of corruption and the consequent coups and regime changes that sometimes result.[65]

Improved investment climate

The enhanced accountability and corporate image mentioned above can culminate in an improved investment climate for a country. As one study explains, "corruption and lack of transparency…can deter foreign invest-ment in the development of natural resources and, thus, can reduce important sources of revenue and the economic development of the host developing countries."[66] Involvement in the EITI, by contrast, can positively affect foreign direct investment in sectors beyond oil, gas, and minerals. The EITI in Nigeria, for example, "played a powerful role in demonstrating a resource-rich nation's commitment to open management of its wealth, thus encouraging investors to view the country as less risky."[67] One high-ranking Nigerian minister recently attributed her country's improved credit rating, improved sovereign debt rating, larger intake of development aid, and improved currency exchange rate to the EITI's ability to "uncover, for public scrutiny, the previously hidden fiscal flows of the oil and gas industry."[68] Indeed, at least two independent studies have found that Nigerian participa-tion in the EITI has encouraged investment and reduced the overall cost of borrowing money.[69]

Community dialogue, and cooperation

The EITI has profited citizens and civil society groups able to receive accu-rate information about the extractive industries sector, whereby they have initiated a dialogue holding governments and companies more account-able, leading to less social conflict. Lack of transparency and the belief that revenues are being maliciously misappropriated or improperly managed can create social divisiveness and fan social tensions that can lead to conflict. By contrast, legal scholars Matthew Genasci and Sarah Pray have found that "with the financial relationship between states and extractive firms laid bare, citizens have the information necessary to exert pressure on government in order to allocate its recourses toward poverty alleviation and develop-ment programs."[70] Karin Lissakers, former President of the Revenue Watch Institute, an NGO, has similarly noted that "the EITI has been instrumental in increasing citizen participation."[71] Thus, the EITI has built trust between host communities, governments, and companies to the point where they can all come peacefully to the table and may forgo more severe tactics such as boycotts or, in extreme cases, violence.[72]

Indeed, one independent assessment of the EITI conducted by Alexandra Gillies and Antoine Heuty from the Revenue Watch Institute found that Candidate and Compliant countries were more likely to redi-rect resources to the poor and that transparency has had a positive impact on the creation of new civic associations. Furthermore, they noted that EITI countries had "less leakage in public expenditures" and a "better public understanding of decision-making." They documented a greater fulfillment of "socioeconomic rights" such as the access to drinkable water or sanitation, better utilization and delivery of public services, and an increased responsiveness on behalf of state planners to public needs among EITI countries.[73]

Transparency as a global norm

The EITI has, lastly, contributed towards the solidification of transparency as a global norm in international law. As proof of this norm, the EITI has already prompted joint agreements emphasizing the importance of transparency between Nigeria and São Tomé e Príncipe, and the Sudanese Government and the Sudan People's Liberation Movement. Following their involvement in the EITI, Nigeria and São Tomé e Príncipe signed an "Abuja Joint Declaration" in 2004 requiring all oil contacts and payments throughout a "Joint Development Zone" encompassing offshore oil reserves in the Gulf of Guinea, thought to hold billions of barrels of oil, to be transparent and made public. The Joint Declaration also requires that all funds received by either government from the Joint Development Zone be monitored and audited externally and independently.[74] Another landmark legal arrangement, inspired by the EITI, was implemented between the Government of the Republic of Sudan and the Sudanese People's Liberation Movement, a rebel group seeking independence, in 2006 before the country of South Sudan was created. This arrangement called for a legal framework to ensure transparency and fairness in revenue management from the oil and gas fields in Southern Sudan.[75]

Challenges

Though the EITI has garnered an impressive list of benefits, it also confronts six challenges related to (1) a limited mandate, (2) its voluntary nature, (3) resistance from the public and private sector, (4) dependence on strong civil society, (5) green-washing, and (6) strategic manipulation.

Limited mandate

One basic challenge is that the EITI focuses only on revenues from the extractive industries in countries rich in resources. This takes a "narrow" view of transparency, as it is only a small part of public sector revenues. Even other aspects of the oil, gas, and mining fuel cycle, such as environmental impact assessments, project siting, or community relocation, are excluded. The EITI, moreover, does not address how those revenues are expended; it merely makes their amounts more precisely known to outside groups. As one commentator criticized:

> [The EITI is problematic because it focuses] on transparency in government oil revenue, or the financial flows between the oil industry and national treasuries, and misses where the corruption is often far worse: in government spending.[76]

Relatedly, until recently about half of EITI countries published aggregate data about revenues but not individual data about particular companies (on the private side) or ministries and departments (on the government side), making it difficult to determine precisely where the money went, though a new EITI standard to be adopted in Sydney in May 2013 will require disaggregated reporting. One critique of the EITI has argued that " the EITI initiative is not only narrow, but it also gives priority to the wrong set of issues in resource-rich countries... since the spread of corruption starts at the early stages involving contracts and procurement, the EITI is introduced too late in the process to have much of an effect."[77]

Voluntary rather than mandatory compliance

Another fundamental weakness of the EITI is that it is "purely a voluntary approach," where governments are "encouraged" but not "required" to adhere to the principles of transparency. This means that only governments and companies committed to integrity and transparency will join.[78] Or, as one legal scholar remarked, "corporations have strong incentives to agree to nondisclosure demands made by resource-rich countries" and "unaccountable governments have equally strong incentives not to change."[79] One survey noted that companies wishing to evade taxes or to quickly maximize profits will opt out of participating in the EITI or will leave countries about to join the EITI.[80] Moreover, for countries that do join voluntarily, there are no sanctions against noncompliance other than rejecting a country's Candidate status – there are no fines, criminal charges, or other penalties. This may create an incentive for corrupt and nontransparent companies and countries to join the EITI in the knowledge that, in a best case scenario, they gain increased prestige and recognition at low cost and, in a worst case scenario, they lose nothing if expelled from the EITI.[81]

As a sign of its limited coverage, although a few dozen countries have participated so far, more than 130 countries around the world produce and extract oil and 86 countries extract significant amounts of coal. Angola, the country whose report from Global Witness inspired the creation of the entire EITI framework, has not joined. China, another major global actor in resource extraction, has also refrained from participating.[82]

Public and private sector resistance

In some circumstances, the EITI can damage companies and communities. Corporate leaders have stated that the voluntary nature of the EITI can put participating companies at a "competitive disadvantage" when they have to disclose information about royalties within the EITI countries they operate in.[83] Other companies have "objected strongly" to the publishing of production-sharing agreements on public websites.[84] Some members of civil society have been harassed and intimidated for participating in the implementation of the EITI, others have seen travel permits and visas denied, and still others have seen legal and procedural obstacles thrown in their way to prevent them from fully participating.[85] In 2006, for example, two members of the civil society coalition involved in promoting the EITI in the Republic of the Congo were arrested and imprisoned.[86]

Dependence on strong civil society

For the EITI to work, it needs strong civil society institutions. Indeed, even for transparency to work effectively, information must become firmly embedded in the everyday decision-making practices of information producers and consumers, creating a transparency "action cycle."[87] Yet in many countries, especially those most prone to corruption, nongovernmental organizations remain disorganized, weak, or even nonexistent. Moreover, the EITI criterion that civil groups have to be "actively engaged as a participant in the design, monitoring and evaluation of this process and contribute towards public debate" can functionally exclude the process from starting in countries until sufficient civil society capacity exists.[88] Even then, Yale Law School scholar Alex Kardon comments that "achieving transparency may not cure the curse where civil society is not strong enough to convert information into accountability."[89] Joseph Bell, one of the experts from Columbia University who helped draft an oil management law for the small island of São Tomé e Príncipe (see Chapter Six), also admits that "transparency cannot [by itself] ensure the responsible use of resource revenues."[90]

An evaluation of EITI in Madagascar, for instance, noted that a lack of civil society meant that powerful mining companies were able to successfully override efforts to achieve "good governance," transparency, and proper engagement with communities.[91] The researchers found that the absence of civil society meant that personal gain from extractive industry contracts was still the largest determinant of which projects go forward, and that, due to pressure from companies and the presence of corruption, national leaders were more concerned with courting investment and maximizing revenues than with transparency.

Green-washing

Some have accused the EITI of enabling companies to perpetuate "green-washing," the "unjustified appropriation of environmental virtue by a company or an industry,"[92] since extractive industries remain by and large socially and environmentally damaging. Oil, natural gas, and coal are still produced by technically complex processes that involve capital-intensive technologies and a host of negative externalities.[93] As one scholar succinctly put it, "few industrial activities have as large an environmental footprint and are capable of wielding as much influence on the wellbeing of a society as a large-scale mine or oil and gas project."[94] It is impossible to construct massive open-pit mines or to build thousands of kilometers of pipelines and processing stations without disturbing communities and ecosystems.

Indeed, some commentators have argued that the impacts of oil on human health, on occupational safety, on deforestation, on water quality, and on air pollution can be somewhat "controlled" but "never eliminated." As Keith Slack from Oxfam writes:

> A contradiction thus exists between commitments to operate responsibly and the actual mechanics of how the industry currently functions. Displacing a community of thousands of people in order to dig a massive pit in a pristine mountain or rainforest area, and piling up 300 meter-high mountains of waste rock that will inevitably begin to leach sulfuric acid into groundwater used by local communities will never be seen as socially responsible by some observers.[95]

Slack concluded that, for these industries, truly operating responsibly amounts to a contradiction in terms, a "mission impossible." In these types of situations, the accession to transparency campaigns such as the EITI may amount to "mere cosmetics, another display of supposed good governance to please the international financial institutions and donor community" without lasting commitments to change or systematic improvements in social welfare or environmental quality.[96] Canadian author and social activist Naomi Klein is even more pithy in her assessment: "With the fossil-fuel industry, wrecking the planet is their business model. It's what they do."[97]

Strategic manipulation

A final challenge, perhaps counter-intuitively, relates to the potential downsides of transparency. The EITI can have a prophylactic effect on oil and gas development, since both governments and companies, knowing that they are being monitored, will significantly change their behavior. Development experts Ivar Kolstad and Arne Wiig explain that:

> A public sector that is to always keep the public informed on all details of its activities will not be very effective in pursuing its activities. In other words, if you keep a diary of everything you do, you won't be doing much. [98]

The EITI has the potential to make negotiations between governments and companies more complex and cumbersome, since parties involved in the process may be more cautious about exchanging information they know will make it into the public sphere. Ironically, transparency can not only make it easier to detect corruption; it can also, in rare instances, identify the relevant officials to approach for bribes and kickbacks.[99]

Conclusion: lessons and implications

The EITI reveals five important things about energy justice. First, it affirms the import and power of information. Merely by enabling the provision of accurate information about oil, gas, and mining revenues, the EITI has helped bridge the schisms that can develop between society, government, and industry. This "opening up of the books" can build trust between stakeholders, hold governments more accountable for billions of dollars of revenue, improve corporate image and the national investment climate, and empower communities – all leading to "greater political and social stability."[100] The old adage that "sunlight is the best disinfectant" comes to mind, and the EITI removes the metaphorical curtain from the window so that the sunlight can enter.

Second, the EITI, largely driven by civil society, confirms that small groups of organized individuals can make a difference. In the case of the EITI, it was motivated by a single NGO, Global Witness, releasing one report about Angola. The EITI similarly implies the necessity of having strong, dedicated leaders and political champions. In the case of the EITI, these included the billionaire George Soros, the British Prime Minister Tony Blair, and Transparency International founder Peter Eigen.

Third, like our Danish case study, the EITI affirms the value of polycentrism and stakeholder participation at varying geographic scales. The EITI involves the use of the EITI Board and Secretariat, a global institution, as well as international NGOs such as Publish What you Pay, Global Witness, Transparency International, and Revenue Watch, in addition to local community groups, national planners, and energy companies.

Fourth, the EITI suggests that, left to their own devices, corporations and corrupt governments need not always race to the bottom to lower standards and perpetuate the resource curse. Instead, the EITI sees 18 governments fully compliant, 19 more considering their candidacy, 60 large energy firms, and more than 900 smaller enterprises participating in a completely voluntary scheme that has the potential to put them at a competitive disadvantage, all in the name of accountability and good governance. Their participation supports the contention that positive norms and codes of conduct can proliferate quickly within an industry.

Fifth, however, the EITI also demonstrates the inescapable social and economic damage from extractive industries, which transparency is unable to fully mitigate. This serves as a harsh reminder that energy systems, particularly fossil fuels, have malicious environmental (and at times social) costs.[101] The EITI tries to ensure that the revenues from their extraction become more accountable, but it does not ensure that the activities themselves become more socially and environmentally sustainable.

6

Prudence and São Tomé e Príncipe's Oil Revenue Management Law

Introduction

The world's four primary energy fuels – oil, natural gas, coal, and uranium – took more than two billion years to accumulate, yet face depletion at a rate that will have them mostly exhausted within one to two centuries, possibly less. Moreover, their production, extraction, and processing have often occurred with devastating consequences for national governments and local communities. After summarizing these concerns, this chapter presents the notion of prudence, the idea that energy resources ought to be maximized for future use, and utilized to better the communities living near them. It then introduces natural resource funds, a mechanism both to moderate the production and extraction of energy resources and to ensure that energy-related revenues serve the public good rather than merely consolidate corporate profit. The natural resource fund from the small island country of São Tomé e Príncipe (STP) offers an excellent case study of how to achieve this balance in practice. National planners there have relied on a combination of an Oil Revenue Management Law, an independent National Petroleum Council, an independent National Petroleum Agency, and a commitment to the EITI (see Chapter Five) to moderate oil and gas production and to ensure that revenues flow democratically to public projects. These policies have generated much-needed government revenue, helped diversify the economy, lowered inflation and rates of poverty, and minimized corruption and the exploitation often associated with oil exploration and production.

Prudence as an energy justice concern

Natural resources serve as the economic bedrock for most developing economies, which still rely on sectors such as mining, agriculture, and forestry to produce the bulk of their income. Natural resources especially account for the largest share of wealth and economic output for least developed countries. For example, almost one-third of GDP in low-income countries is produced directly from natural resources compared with less than 13 percent for middle-income countries.[1] When looked at historically, however, natural resource wealth is declining. In many regions, such as Africa, it has dropped as a ratio of national exports by more than 30 percent from 1970 to 2005.[2]

One of the resources most prone to rapid depletion has been crude oil. To date the world has consumed about one trillion barrels of oil. Although oil production has occurred since the nineteenth century, 99.5 percent has happened within the last 60 years. Today we consume 30 *billion* barrels of oil per year, and future demand for oil is projected to grow at more than twice the historic rate since 1980. The world's known 1.2 trillion barrels of oil reserves are concentrated in a handful of countries around the world, and would last only a few decades at current rates of consumption. From 2002 to 2011, the oil industry's annual spending on exploration and production has increased by a factor of four, yet oil production has risen by only 12 percent.[3]

The distribution of other conventional energy resources, such as coal, natural gas, and uranium, is equally consolidated. Figure 6.1 provides a graphic illustration of energy reserves by country. Strikingly, 80 percent of the world's oil can be found in nine countries that have only five percent of the world's population and 5 percent of its GDP; 80 percent of the world's natural gas is in 13 countries with 12 percent of the population and 26

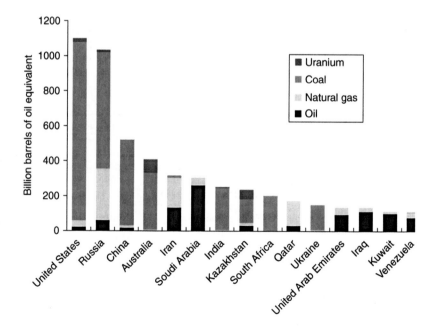

Figure 6.1   Major global energy reserves for the top 15 countries, 2008

Source: Adapted from Brown and Sovacool. Excludes unconventional resources such as shale gas, oil tarsands, methane hydrates, and thorium.

percent of GDP; and 80 percent of the world's coal is in six countries (though these countries have 45 percent of the population and 46 percent of GDP). Many of the same countries are among the six that control more than 80 percent of global uranium resources.

Although these reserves may seem vast, a growing worldwide demand for electricity and mobility threatens to exhaust them relatively soon. Put another way, the world is transitioning from a position of abundant fossil energy supplies to a largely resource-constrained supply future. World energy demand is expected to expand by 45 percent between now and 2030, and by more than 300 percent by the end of the century.[4] If levels of production were to remain constant worldwide, Table 6.1 illustrates that known coal reserves would exhaust themselves within 137 years, and petroleum and natural gas reserves (excluding unconventional reserves such as tarsands or shale gas) would exhaust themselves in the next half-century. If rates of production increased to keep up with growing demand, particularly in the rapidly developing BRIC countries (Brazil, Russia, India, and China), known fossil fuel reserves would be depleted even more rapidly.

Additionally, Table 6.1 shows that, if the world were to maintain its generation of nuclear power at 2004 levels, identified uranium resources would run out in 85 years. The International Atomic Energy Agency (IAEA), after collecting information on 582 uranium mines and deposits worldwide, expected primary supply to cover 42 percent of demand for uranium in 2000, but acknowledges that the number will drop to between four percent and six percent of supply in 2025, as low-cost ores are expended and countries are forced to explore harder to reach and more expensive sites.[5] But here lies a

Table 6.1   Life expectancy of proven fossil fuel and uranium resources, 2008

	Proven Reserves	Current Production	Life Expectancy (Years)		
			0% Annual Production Growth Rate	2.5% Production Growth Rate	5% Production Growth Rate
Coal	930,400 million short tons	6,807 million short tons	137	60	42
Natural Gas	6,189 trillion cubic feet	104.0 trillion cubic feet	60	37	28
Petroleum	1317 billion barrels	30.560 billion barrels	43	29	23
Uranium	4,743,000 tons (at $130/kgU)	40,260 tons	85	46	33

Source: Adapted from Brown and Sovacool.

conundrum: the IAEA calculates that secondary supply can only contribute 8–11 percent of world demand. "As we look to the future, presently known resources fall short of demand," the IAEA stated in 2001, and "it will become necessary to rely on very high cost conventional or unconventional resources to meet demand as the lower cost known resources are exhausted."[6]

It is thus unclear how long the world can produce enough oil (and even coal, natural gas, and uranium) to meet growing global demand. M. King Hubbert, the famous geophysicist who predicted that American oil production would peak about 1970, often remarked that it would be incredibly difficult for people living now, accustomed to exponential growth in energy consumption, to assess the transitory nature of fossil fuels – to comprehend just how fast we're using up resources that have taken millions of years to form. Hubbert argued that proper reflection could happen only if one looked at a timescale of millennia.[7] On such a scale, Hubbert thought that the complete cycle of the world's exploitation of fossil fuels would encompass perhaps little more than one thousand years, with the principal segment of this cycle covering about three centuries, shown in Figure 6.2.[8]

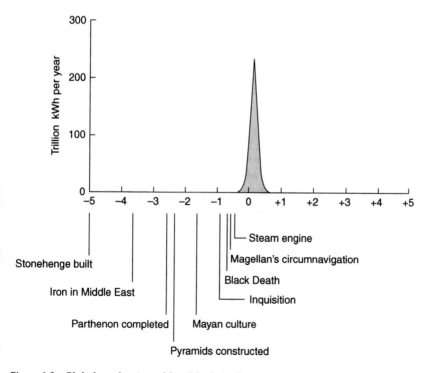

Figure 6.2 Global production of fossil fuels (trillion kWh)
Source: Adapted from Emirates Center for Strategic Studies and Research.

Not only are present generations using all of the resources, they are doing so in ways that spend mineral revenues quickly and poorly. As mineral economists Marcatan Humphreys and Martin E. Sandbu explain:

> A crucial reason for the inability of policymakers to save windfall revenues for the future is that they risk losing control over how money will be spent at a later time. If a new government can come to power and dramatically alter the way the money is spent, an incumbent policy maker has an incentive to spend more money now, even if he himself would also prefer to smooth spending out over future time periods.[9]

The result has been rampant and widespread accounts of corruption within the energy sector. One 2006 survey conducted by the Control Risks Group found that 23 percent of managers stated their company "had lost business" due to bribes, and a 2008 Ernst & Young survey found that one-quarter of executives reported an "incident of bribery within the past two years."[10]

Various types of corruption regarding revenues from oil exploration and production – the looting of oil revenues by well-placed military and political representatives, kickbacks, bribes, illegal commissions, and illicit oil-for-arms deals – routinely occur.[11] In Iran, for example, $11 billion in oil revenues "vanished" over the course of nine months in 2010.[12] In Angola, the second largest oil producer in Africa after Nigeria, investigators discovered in December 2000 that $4 billion in oil funds was missing, likely spent illegally on kickbacks to government officials and the purchase of military arms – these missing funds represented a staggering one-third of all state income in Angola.[13] In Congo-Brazzaville, the head of its state-owned oil company, Société nationale des pétroles du Congo, was indicted for illegally taking as much as $4.2 million of company revenues per year and depositing them into his personal account.[14] In Kazakhstan, $20 million was funneled to the president as payments to a "secret fund" to obtain licenses for Mobil to explore petroleum in the Tengiz oilfield.[15] Enthralled with prospects of the Chad-Cameroon oil pipeline, the Chadian government spent $4.5 million of its $25 million "signing bonus" on weapons to start a civil war. Once oil began flowing through the pipeline in 2000, the President purchased a private airplane.[16]

Nigeria deserves special mention, given the sheer magnitude of its corruption from oil revenues. One independent Nigerian audit revealed $540 million "missing" from $1.6 billion in advance payments to oil companies to develop new fields, in addition to 3.1 million barrels of oil that remain unaccounted for, along with the disappearance of $3.8 billion in dividends from natural gas.[17] The administration of Babangida, who ruled between 1985 and 1993, is reputed to have stolen $12.2 billion in oil revenues, and General Abacha's regime is reputed to have embezzled as much as $3 billion.[18] Collectively, since independence in 1960, it is estimated that between $300 and $400 *billion* of oil revenue has been stolen or misspent by corrupt Nigerian government officials – an amount of money approaching all the Western aid received by the entire African continent over the course of those years.[19]

Such corruption is not always at the hands of national politicians. In some cases, it occurs within industrialized countries. In France, for example, the government-linked oil company Elf Gabon propped up various African "oil states" to generate revenues, or an "oily offshore slush fund," that they could tap to funnel money into French political parties seeking reelection.[20]

Furthermore, it's not just governments that are at risk of corruption. Even veritable intergovernmental organizations charged with preserving human rights, such as the United Nations, are susceptible. In 2002 and 2003, it was revealed that the United Nations Oil-for-Food Program – established in 1995 to enable Iraq to sell oil on the world market in exchange for food, medicine, and humanitarian aid – enabled Saddam Hussein and the Iraqi regime to generate $10.1 billion in illegal revenues, including $5.7 billion of smuggling proceeds and $4.4 billion in illicit surcharges on oil products. An independent inquiry by the US Government Accountability Office documented more than 380 illegal acts of corruption, with some of the revenues going directly into the pockets of UN officials managing the program.[21]

At whatever level it occurs, corruption has a corrosive effect on social stability and economic development. As corruption expert Michael Johnston, Charles A. Dana Professor of Political Science at Colgate University, put it:

> Corruption undermines the quality of government as well as its accountability to society at large. It disrupts fair and openly competitive economic and political processes, generally benefiting the "haves" – that is, the well-connected and their clients – at the expense of the "have-nots," whose opportunities often depend more upon fair procedures and dependable rules. It can offer sizeable material gains while draining away resources and weakening the property and political rights of society at large.[22]

The ability of oil, gas, and mineral production to facilitate corruption serves as a powerful causal contributor to poverty and economic stagnation.

To address the twin problems of resource exhaustion and corruption, the principle of prudence calls for energy resources to be depleted more sustainably and with revenues that produce a positive, rather than negative, impact on society. A truly prudent energy strategy is encapsulated in the following statement from the philosopher Douglas MacLean, who wrote that:

> None of the problems we confront in forming an energy policy is more important or more basic than finding a satisfactory balance between securing the resources to meet our own current needs and conserving and

protecting these resources and the environment for the use of generations that will succeed us.[23]

This is consistent with the original definition of sustainability as set forth in *Our Common Future*, the report of the World Commission on Environment and Development, also known as the Brundtland Report. According to this report, sustainable development "meets the needs of the present without compromising the ability of future generations to meet their own needs."[24]

Georgetown law professor Edith Brown Weiss argues that prudence can be framed to be about conserving options, quality, and access for future generations. As she notes, each generation should be required to conserve the diversity of the natural and cultural resource base, so it does not restrict options for future generations, their ability to solve their problems, and their capacity to satisfy their own values (something known as "conservation of options"). Each generation should maintain the quality of the planet passed on in no worse condition than they received it (something known as "conservation of quality"). Each generation should provide its members equitable access to the legacy from past generations, conserving this access to future generations, passing it along in a sort of intergenerational trust (something known as "conservation of access"). These notions do overlap with other justice concerns – notably those of due process and human rights (Chapter Four), transparency (Chapter Five), intergenerational equity (Chapter Seven), and intragenerational equity (Chapter Eight). Yet prudence is distinct because it suggests that we prudently use natural resources to leave enough for others.

Sovereign Wealth Funds – sometimes called National Wealth Funds or Natural Resource Funds (NRFs) – exist so that the countries can convert the proceeds of natural resource extraction into financial assets.[25] NRFs date back to the 1950s and 1960s, when Kuwait established a "General Resource Fund" for crude oil and Kiribati created its "Revenue Equalization Reserve Fund" for phosphate revenues, followed by the "best practice" of Norway with its Oil Fund in 1967. As of 2008, 33 countries around the world had implemented NRFs, with roughly $3 trillion worth of investment, as shown in Table 6.2.[26] Such funds have become so important to both national development and the global economy that analysts have described them as "lifelines"[27] and "stabilizing influences" for financial markets.[28]

NRFs, when used properly, offer an outstanding mechanism for achieving prudence in practice. As Humphreys and Sandbu note, "the stated purpose of [NRFs] is to facilitate the accumulation of large, volatile, and temporary revenues when times are good; stabilize public spending; and finance public spending when natural resource revenues are no longer flowing in."[29] NRFs essentially manage revenues from energy and other natural resources as a "trustee for the benefit of citizens, including future generations," avoiding boom and bust cycles, reducing corruption and rent-seeking activity, improving governance and transparency, and establishing links between the

Table 6.2 Global natural resource funds, 2008

Country	Current Name	Date Established	Source of Funds	Size (billions of US$)
Algeria	Revenue Regulation Fund	2000	Natural resources	47
Azerbaijan	State Oil Fund of the Republic of Azerbaijan	1999	Natural resources	2
Botswana	Pula Fund	1993	Natural resources	7
Brunei	Brunei Investment Agency Alberta	1983	Natural resources	35
Canada	Heritage Savings Trust Fund	1976	Natural resources	17
Chile	Economic and Social Stabilization Fund	2006	Natural resources	15
China	China Investment Corporation	2007	Foreign exchange reserves	200
	Shanghai Financial Holdings	2007	Fiscal surpluses	1
Gabon	Fund for Future Generations	1998	Natural resources	0.4
Hong Kong	Exchange Fund Investment Portfolio	1993	Foreign exchange reserves, fiscal surpluses	139
Iran	Oil Stabilization Fund	2000	Natural resources	10
Kazakhstan	National Fund for the Republic of Kazakhstan	2000	Natural resources	23
Kiribati	Revenue Equalization Reserve Fund	1956	Natural resources	1
Korea	Korea Investment Corporation	2005	Foreign exchange reserves	30
Kuwait	Kuwait Investment Authority	1953	Natural resources	213
Libya	Libyan Investment Authority	2006	Natural resources	50
Malaysia	Khazanah Nasional	1993	Fiscal surpluses	18
Mexico	Oil Income Stabilization Fund	2000	Natural resources	5
Nigeria	Excess Crude Account	2003	Natural resources	17
Norway	Government Pension Fund – Global	1990	Natural resources	375
Oman	State General Reserve Fund	1980	Natural resources	13

Continued

Table 6.2   Continued

Country	Current Name	Date Established	Source of Funds	Size (billions of US$)
Qatar	Qatar Investment Authority	2005	Natural resources	60
Russia	National Wealth Fund	2008	Natural resources	32
	Reserve Fund	2008	Natural resources	128
São Tomé and Príncipe	National Oil Account	2004	Natural resources	0.02
Saudi Arabia	Saudi Arabian Monetary Agency	1952	Natural resources	270
Singapore	Government of Singapore Investment Corporation	1981	Foreign exchange reserves, fiscal surpluses, employee contributions	275
	Temasek Holdings	1974	Government enterprises	110
Sudan	Oil Revenue Stabilization Account	2002	Natural resources	0.1
Timor-Leste	Petroleum Fund	2005	Natural resources	2
Trinidad and Tobago	Heritage and Stabilization Fund	2007	Natural resources	2
United Arab Emirates	Emirates Investment Authority	2007	Natural resources	.
United Arab Emirates (Abu Dhabi)	Abu Dhabi Investment Authority and Council	1976	Natural resources	740
	International Petroleum Investment Company	1984	Natural resources	12
	Mubadala Development Company	2002	Natural resources	10
United Arab Emirates (Dubai)	DIFC Investments	2006	Natural resources	
	Dubai International Capital	2004	Natural resources	13
	Investment Corporation of Dubai	2006	Natural resources	82
	Istithmar	2003	Natural resources	12

Continued

Table 6.2   Continued

Country	Current Name	Date Established	Source of Funds	Size (billions of US$)
United States	Alaska Permanent Fund	1976	Natural resources	37
	Permanent Mineral Trust Fund (Wyoming)	1974	Natural resources	4
	Severance Tax Permanent Fund (New Mexico)	1973	Natural resources	5
Venezuela	Macroeconomic Stabilization Fund	1998	Natural resources	1
	National Development Fund	2005	Natural resources	21
Total				3,035

Source: Adapted from Truman.

extractive industries and ordinary citizens.[30] NRFs can create an "endowment effect" whereby citizens require the government to justify how it spends revenues and resources; an "information effect" whereby citizens better understand government procurement; and an "income effect" whereby communities benefit from increased government spending on programs that improve standards of living.[31] Professor of Law Abdullah Al Faruque writes, therefore, that "national resource funds can obviously ... facilitate the wellbeing of the vast majority of the people who are hitherto deprived from the fruits of exploitation of natural resources."[32]

Case study: São Tomé e Príncipe's Oil Revenue Management Law

STP's natural resource fund involved the creation of an independent National Petroleum Council and National Petroleum Agency, a groundbreaking Oil Revenue Management Law, and a commitment to the principles of transparency stipulated by the EITI (see Chapter Five for more on that element). Taken together, these efforts have been called "the model for other developing countries who find themselves suddenly in control of vast resource wealth,"[33] hailed for "leading the way to greater transparency in the natural resource industry,"[34] and praised for demonstrating "best practice in the governance of oil revenues."[35] Even the usually somber International Monetary Fund (IMF) lauded STP's policies for their "sound fiscal properties" and for "ensuring transparent and accountable management of oil resources."[36]

History

For those who have never visited, the Democratic Republic of São Tomé e Príncipe (hereafter STP) is an archipelago of roughly one thousand square kilometers in the Gulf of Guinea, approximately 300 kilometers from the coast of Equatorial Guinea. With a population of 160,000, it is West Africa's smallest democracy, Africa's second smallest country after the Seychelles, and the third smallest economy in the world. To put its smallness in perspective, the country is about five times the size of Washington, DC. STP is also unique in that it is an "invented society" that traces its roots to a Portuguese colony populated by white Portuguese colonists and West African slaves.[37]

Due to its favorable location, STP was central to a Gulf of Guinea oil "bonanza" in the late 1980s and early 1990s when petroleum geologists estimated that the Gulf held more than 50 billion barrels of oil, or about five percent of the world's proven oil reserves. Some of this oil falls into two areas under the sovereignty of STP: within the country's Exclusive Economic Zone (EEZ) stipulated by the United Nations Convention on the Law of the Sea, and within an offshore Joint Development Zone (JDZ) established with Nigeria in 2001, shown in Figure 6.3. Given that it crosses both countries' EEZs, the JDZ offers Nigeria 60 percent of the benefits and obligations arising from development activities within the Zone and STP 40 percent. Block 1 of the JDZ was thought to contain between 1 and 1.5 billion barrels of oil, enough to convince a consortium of major companies involving Chevron, ExxonMobil, and the Nigerian company Dangote Energy Equity Resources to pay $123 million to STP for a production-sharing contract and signature bonuses, with "very promising prospects for commercially viable oil discoveries" in other areas.[38] Under STP law, specifically the General Law on Petroleum Exploration and Exploitation passed in August 2000, all hydrocarbon reserves belong to the state. This "bonanza" catapulted the STP government and the entire Gulf "from strategic neglect into geopolitical stardom."[39]

At the time of the major oil discoveries in the late 1990s, however, STP was almost entirely dependent on foreign aid, notorious for somewhat humorous cases of petty corruption and, due to lack of money, government neglect and a highly nondiversified economy. For instance, in 1995 STP received $84 million in overseas development assistance, about *twice* the country's actual GDP, and aid represented 97 percent of government revenue, making it the largest recipient of donor aid per capita in the world. It also had the highest debt to GDP ratio of any country in the world. At the same time, however, government officials were notorious for their petty corruption, with some of this aid functioning "as an illegal source of finance for people within, or close to, the government and the presidency."[40] Another academic article joked that "while in opposition, politicians denounce the corruption, and while in office, they practice it."[41] One infamous scandal involved a senior minister billing the government $48,780 for a "pack of cigarettes." Even STP's

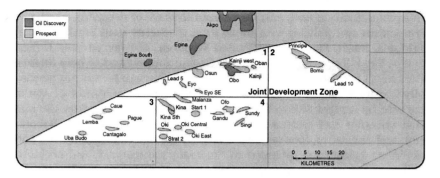

Figure 6.3 The São Tomé e Príncipe-Nigerian Joint Development Zone in the Gulf of Guinea

Source: Adapted from the Joint Development Authority.

former President Miguel Trovoada admitted to having received $100,000 in donations from oil companies towards his election campaign, though this was certainly not illegal (and occurs in places like Europe and the United States as well).[42]

STP faced economic difficulties during this time, and it was incredibly expensive to import virtually everything STP citizens needed for modern life, because there was no economy of scale, with one scholar commenting that "in STP it's not worth building even a shoe factory to produce shoes for a market of only 150,000 customers."[43] STP therefore had to import every light bulb, every car battery, every computer, every car part, every car, every gallon of gasoline, every fork, every plate, every curtain, every shoe, and so on. In STP's case these imports had to come from vast distances, since neighboring countries also produced little and imported almost everything from Europe. STP's geographical isolation made it a difficult vacation destination because it wasn't near tourism markets, and had only one flight a week linking it to Europe. Political scientists and economists have consequently labeled STP an "unviable state" during this period.[44]

Cognizant of this state of affairs, after being elected in 2001, STP President Fradique Menezes is reputed to have called the Columbia University sustainable development guru Jeffrey Sachs in 2001, saying that "we've found some oil and the sharks are swimming around us now...I'd like some help to manage this properly." At that time, major multinational oil companies assumed STP would have oil resources similar to all the other Gulf of Guinea countries, and wanted access, and President de Menezes knew his country had neither the expertise to deal with the oil companies nor a potential oil economy. He wanted to ensure that STP did not go the way of Nigeria, Gabon, Angola, and Equatorial Guinea, which earn billions of dollars each year but whose people continue to live in squalor. Sachs agreed, and then began the task of designing a series of regulations and policies with input from specialists from Mexico, India, Ethiopia, Brazil, Portugal, the United Kingdom, Norway, and the United States. The billionaire philanthropist George Soros agreed to help finance the project through his Open Society Institute, though he was not personally involved. The STP President sought out a former governor of Alaska because he wanted STP to have a Permanent Fund modeled on the one that governor had created for Alaska. The end result of this collaboration and discussion, four separate acts implemented throughout 2004 and described below, was intended to "increase transparency and accountability in the government's use of oil revenue" and protect STP from the "frequent destabilization of an economy following a country's discovery of natural resources."[45]

The first act emerging from this process, Decree Law No. 3/2004, implemented in June 2004, created a National Petroleum Council. Comprised of 15 members, including the President, the Prime Minister, and representatives of civil society, the NPC sets national energy policies.

The second of these acts, Law No. 5/2004 implemented in June 2004, established a National Petroleum Agency (ANP). The ANP is in charge of regulating oil and gas exploration and production according to the policies set by the NPC.

The third of these acts, the Abuja Joint Declaration, signed by the Presidents of Nigeria and São Tomé and Príncipe in June of 2004, established transparency guidelines for all operations within the JDZ. It attempts to adhere to the principles of the EITI (see Chapter Five for more on that front).

The fourth of these acts, the Oil Revenue Management Law (ORML), remains the centerpiece and the most important. Passed unanimously by the Santomean National Assembly in December 2004 after a breathtaking 55 "town hall meetings," it regulates the "payments, management, use, and oversight" of revenues arising from oil operations throughout STP and within the JDZ, revenues that are then placed in a National Oil Account and a Permanent Oil Fund.

The ORML possesses a number of unique attributes. Wary of the potential for oil revenues to encourage corruption, the ORML created a national oil fund, formally known as the "National Oil Account," held by an international and independent custodial bank, the New York Federal Reserve. The ORML mandated that all oil payments be made directly into the fund, bypassing senior ministers and politicians, and demanded that there can be only one annual transfer from the National Oil Account to the government budget. Moreover, that transfer requires the signature of four separate officials, including the President and the Prime Minister. The maximum amount of that transfer is determined according to a fixed formula based on the late economist Milton Friedman's "permanent income hypothesis," which argued that government spending each year should be limited to the real return on the amount that equals financial assets already accumulated plus the net present value of oil reserves still in the ground. Over time, the mechanism attempts to build a financial asset of equal value to the reserves that have been depleted, achieving intergenerational equity.[46] Outflows cannot legally exceed the highest amount of oil production that can be sustained in perpetuity (except for a beginning transitional period).

However, the ORML also set limits on withdrawals from the National Oil Account so that a significant amount of revenues accrue to a sub-account known as the "Permanent Fund for Future Generations" or the "Permanent Oil Fund," which cannot be spent now and forms a "national endowment" to "foster development even after oil resources have been exhausted." Furthermore, annual spending from the National Oil Account must be spent in accordance with the priorities enshrined within the country's Poverty Reduction Strategy, and ten percent given directly to local governments. The ORML requires competitive tenders for all oil and gas contracts and also makes those contracts public.[47] To avoid inflation, all oil revenues are held in foreign denominated currencies; to mitigate corruption, revenues must match deposits precisely.

To ensure compliance, yet another entirely separate institution was created, the Petroleum Oversight Commission. This Commission, including members of government, the National Assembly, opposition parties, and nongovernmental organizations, has "significant investigatory and administrative powers." The ORML requires that each year the National Assembly discusses in plenary sessions the state of oil and gas production and the status of the National Oil Account. Lastly, it has severe penalties for noncompliance, including strict fines as well as criminal sanctions, including imprisonment.[48]

Benefits

Although full-scale oil production in the JDZ has yet to commence, STP's concerted efforts to promote transparency and save for future generations have already (1) enhanced government revenues, (2) diversified the economy, (3) improved macroeconomic indicators related to poverty and inflation, and (4) lessened corruption and protected sovereignty.

Enhanced government revenue

The ORML has ensured that millions of dollars have already entered the Permanent Oil Fund and National Oil Account, and it set the way for the amounts to rise to billions of dollars. Existing revenues have so far come from signature bonuses and royalties on exploration rather than production, but the IMF reports $116 million in expected revenues from 2008 to 2015 – figures summarized in Figure 6.4.[49]

Once oil production begins, the amount of government revenues and deposits into the Permanent Oil Fund will increase astronomically.

Conservative baseline discoveries within Block 1 of the JDZ – merely one oil field – suggest at least 500 million equivalent barrels of oil in reserves, an amount equal to about 70,000 barrels per day for 20 years, as shown in Figure 6.5, of which 28,000 barrels per day would belong to STP (under the Abuja Joint Declaration). Presuming these barrels are worth a paltry $30 per barrel – about one-third the cost of oil today – by 2032 the National Oil Account would see $91.9 million in annual revenue and the Permanent Oil Fund would stabilize at slightly above $3 billion, or 46 times the country's GDP in 2006, numbers summarized in Table 6.3.[50] If oil reserves were twice as high as assumed, the amount would be $6 billion, and if the price of oil stayed close to today's rate of $100 per barrel, the amount would surpass $20 billion.

Greater economic diversification

The increased revenue and savings into the Permanent Oil Fund from exploration and production not only strengthens government spending; it is also diversifying the STP economy. Before oil exploration began, the

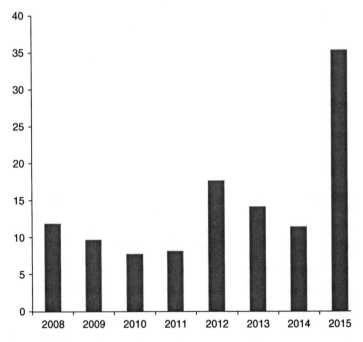

Figure 6.4 Revenues in São Tomé e Príncipe National Oil Account, 2008–2015 (millions of dollars)

Source: International Monetary Fund.

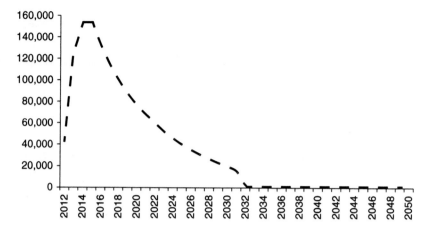

Figure 6.5   Oil production profile for Block One of the Joint Development Zone
Source: International Monetary Fund.

Table 6.3   Projected São Tomé e Príncipe oil flows (in millions of $2006), 2012–2050

	Oil Receipts	Total	Annual Funding for Public Budget			Permanent Oil Fund	
			From annual production	From Permanent Fund	Percent of non-oil GDP	Balance	Percent of non-oil GDP
2012	25.9	0.0	0.0	0.0	0.0	0.0	0.0
2013	90.7	13.0	13.0	0.0	11.3	0.0	0.0
2014	306.6	59.0	59.0	0.0	47.3	0.0	0.0
2015	396.1	73.3	73.3	0.0	54.4	128.2	95.1
2016	364.6	75.9	72.0	3.8	52.1	414.6	284.7
2017	333.6	78.2	65.8	12.4	49.8	739.9	470.5
2018	300.5	80.4	58.2	22.2	47.3	1,042.0	613.5
2019	268.1	82.3	51.1	31.3	44.9	1,318.2	718.7
2020	237.6	84.0	44.5	39.5	42.4	1,567.3	791.2
2021	209.3	85.5	38.5	47.0	40.0	1,790.0	836.6
2022	183.5	86.7	33.0	53.7	37.5	1,987.7	860.2
2023	164.8	87.8	28.2	59.6	35.2	2,162.5	866.5
2024	140.1	88.7	23.8	64.9	32.9	2,318.5	860.3
2025	122.1	89.5	19.9	69.6	30.7	2,456.1	843.8
2026	106.2	90.1	16.4	73.7	28.7	2,575.2	819.2
2027	92.1	90.6	13.4	77.3	26.7	2,679.8	789.3
2028	79.8	91.0	10.6	80.4	24.8	2,771.6	755.9
2029	69.1	91.3	8.2	83.1	23.1	2,852.3	720.3
2030	59.6	91.6	6.0	85.6	21.4	2,923.2	683.5

Continued

Table 6.3   Continued

Oil Receipts		Total	Annual Funding for Public Budget			Permanent Oil Fund	
			From annual production	From Permanent Fund	Percent of non-oil GDP	Balance	Percent of non-oil GDP
2031	47.0	91.7	4.0	87.7	19.9	2,985.6	646.4
2032	0.0	91.8	2.3	89.6	18.4	3,038.5	609.1
2033	0.0	91.9	0.7	91.2	17.1	3,062.7	568.5
2034	0.0	91.9	0.0	91.9	15.8	3,063.2	526.5
2035	0.0	91.9	0.0	91.9	14.6	3,063.2	487.5
2036	0.0	91.9	0.0	91.9	13.5	3,063.2	451.3
2037	0.0	91.9	0.0	91.9	12.5	3,063.2	417.9
2038	0.0	91.9	0.0	91.9	11.6	3,063.2	387.0
2039	0.0	91.9	0.0	91.9	10.7	3,063.2	358.3
2040	0.0	91.9	0.0	91.9	10.0	3,063.2	331.8
2041	0.0	91.9	0.0	91.9	9.2	3,063.2	307.2
2042	0.0	91.9	0.0	91.9	8.5	3,063.2	284.4
2043	0.0	91.9	0.0	91.9	7.9	3,063.2	263.4
2044	0.0	91.9	0.0	91.9	7.3	3,063.2	243.8
2045	0.0	91.9	0.0	91.9	6.8	3,063.2	225.8
2046	0.0	91.9	0.0	91.9	6.3	3,063.2	209.1
2047	0.0	91.9	0.0	91.9	5.8	3,063.2	193.6
2048	0.0	91.9	0.0	91.9	5.4	3,063.2	179.2
2049	0.0	91.9	0.0	91.9	5.0	3,063.2	166.0
2050	0.0	91.9	0.0	91.9	4.6	3,063.2	153.7

Source: Adapted from International Monetary Fund.

country's economy was highly undiversified, with cocoa production as the main commercial crop producing less than $4 million per year, and income from tourism marginal. Due to its sheer lack of resources and high rates of debt, the STP economy has featured numerous "informal" and "creative" enterprises. Efforts in the past, for instance, have included offering black market passports, licensing flags of convenience for international shipping companies, producing commemorative postal stamps featuring Marilyn Monroe and the Beatles, and even government-endorsed X-rated telephone calling centers.[51] The National Oil Account, in contrast, has already generated money for "real" money-making government expenditures directed at raising agricultural productivity, improving food security, building basic infrastructure such as hospitals and roads, and promoting a niche tourism market.[52] The IMF has confirmed that, due largely to the National Oil Account, construction and tourism have started to represent more significant shares of GDP.[53]

Macroeconomic stability

Though this is not entirely the result of the ORML and other transparency efforts, their implementation has contributed to a climate of trust and investor confidence, substantially improving STP's macroeconomic outlook. In 2003, the year before the ORML was passed, GDP per capita in STP was a meager $300.[54] Now, in 2012, it is $2,300. The country has a harmonized credit rating from the World Bank and African Development Bank that has left the realm of "fragile states."[55] Inflation has declined every year since 2008 – with a 14 percent drop in 2010 – and real GDP has grown five percent in 2011, reflecting what the IMF has called a "rebound in foreign direct investment and other private sector capital inflows supported by ... prudent fiscal policy."[56] The IMF has also noted that:

> The strategy adopted by São Toméan authorities remains prudent, and focuses on mobilizing domestic revenue to create additional fiscal space for infrastructure and pro-poor spending, while keeping the domestic primary deficit in line with non-debt creating financing.[57]

The IMF also applauded the government for reaching a balanced budget, for paying its fuel tax arrears on time, for reconciling unpaid bills with its electricity and water utilities, and for strengthening the administration and collection of taxes.

Less corruption, more sovereignty

Though difficult to measure, the ORML has improved national governance by enhancing sovereignty, reducing exploitation, and lowering corruption. Before, transnational oil companies were renowned for "swooping in" and "controlling negotiations with the unsophisticated Sao Tomeans."[58] Even the World Bank described the contractual terms between Nigerian and American oil companies and the STP government in the 1990s as "unprecedented" in the amount of resources they awarded foreign entities. In both a tragic and a comic act of exploitation, Nigeria impounded and refused to release STP's entire navy in order to gain more favorable oil contracts. The process of creating the ORML helped counter these acts of exploitation by setting strict limitations on how oil revenues can be generated and by building the capacity of STP planners through their dialogues with experts at Columbia University, the IMF, and the Open Society Institute.

In addition, the NRF has been credited with lowering corruption and improving government accountability. The African Development Fund recently noted that, during the period of implementing the ORML and other components of its NRF, "Sao Tome and Principe has made great strides toward developing its democratic institutions and further guaranteeing the civil and human rights of its citizens."[59] The 2010 edition of the Ibrahim index of African Governance ranked the country 11th out

Table 6.4 Natural resource fund/sovereign wealth fund performance scores for selected developing countries, 2008

Country	Name	Structure	Governance	Transparency	Behavior	Total
Thailand	Government Pension Fund	100	100	88	42	84
Timor-Leste	Petroleum Fund for Timor-Leste	100	40	96	50	80
Azerbaijan	State Oil Fund of the Republic of Azerbaijan	88	60	89	50	77
China	National Social Security Fund	100	40	82	67	77
Chile	Economic and Social Stabilization Fund	94	60	86	17	71
Chile	Pension Reserve Fund	94	60	86	17	71
Kazakhstan	National Fund for the Republic of Kazakhstan	88	60	64	33	64
Botswana	Pula Fund	69	60	54	33	55
Trinidad and Tobago	Heritage and Stabilization Fund	100	60	46	0	53
São Tomé and Príncipe	National Oil Account	100	60	29	17	48
Kuwait	Kuwait Investment Authority	75	80	41	0	48
Mexico	Oil Income Stabilization Fund	69	20	43	50	47
Singapore	Temasek Holdings	50	50	61	0	45
Singapore	Government of Singapore Investment Corporation	63	40	39	17	41
Malaysia	Khazanah Nasional	44	50	46	0	38

China	China Investment Corporation	50	50	14	17	29
Kiribati	Revenue Equalization Reserve Fund	69	60	7	0	29
Algeria	Revenue Regulation Fund	56	40	11	17	27
Nigeria	Excess Crude Account	50	30	14	17	26
Iran	Oil Stabilization Fund	50	20	18	0	23
Venezuela	Macroeconomic Stabilization Fund	50	0	18	17	23
Venezuela	National Development Fund	38	0	27	0	20
Oman	State General Reserve Fund	50	0	18	0	20
Sudan	Oil Revenue Stabilization Account	56	0	14	0	20
Brunei Darussalam	Brunei Investment Agency	31	0	25	0	18
United Arab Emirates (Abu Dhabi)	Mubadala Development Company	44	10	7	0	15
United Arab Emirates (Dubai)	Istithmar World	38	10	7	0	14
Qatar	Qatar Investment Authority	34	0	2	0	9
United Arab Emirates (Abu Dhabi)	Abu Dhabi Investment Authority and Council	25	0	4	8	9

Source: Adapted from Truman.

of 53 countries in terms of their protection of civil liberties and political rights.[60] Moreover, an independent scorecard of "best practices" for Sovereign Wealth Funds (inclusive of NRFs), which rated them according to (1) structure, (2) governance, (3) accountability and transparency, and (4) behavior, revealed that STP is in the top ten of all performers when industrialized countries such as Canada, Japan, Norway, and the United States are excluded. As Table 6.4 depicts, this places STP ahead of countries such as Brunei, Kuwait, Nigeria, Qatar, Singapore, and the United Arab Emirates.[61]

Challenges

In the implementation of its ORML, STP has still confronted obstacles relating to (1) technical difficulties exploiting oil reserves, (2) continuing economic turmoil, and (3) opacity and corruption.

Technical difficulties producing oil

Though STP and the Gulf of Guinea hold ample amounts of theoretical oil, exploring and producing it is prone to technical difficulties. Attempts at oil production date back to the early 1970s, when the Portuguese colonial bureaucracy signed a concession agreement with the American company Ball & Collins that never materialized, followed by "several failed attempts" to start petroleum exploration in the 1980s and early 1990s. These efforts all collapsed due to a combination of financial losses, maritime security issues, and expensive offshore drilling operations. One of these issues, complicated drilling and expensive production costs, extends to this decade. Oil within the JDZ is hard to produce, since it is "ultra-deep," given the distance from the tongue that comes out from the Niger River Delta. Existing technology cannot reach it in a cost-effective manner, with production costs currently estimated at being viable when the price of crude oil surpasses $200 per barrel.[62] As such, oil production in the JDZ is as much as five years behind schedule due to the need for complicated horizontal drilling and uncertainty over future oil prices.[63]

Continuing economic malaise

Though improving, STP is unable to support itself economically even with the proceeds from the National Oil Account, and without actual oil production it's perhaps absurd to think STP could support itself on signature bonuses alone. While the country's GDP has improved remarkably, its agricultural output has remained stagnant and roughly half of the population (54 percent) remains below the international poverty line. One expert identified two existing classes of people: the "poor" and the "extremely poor."[64] In 2008, despite the signing of bonuses from oil exploration, the government could fund only 20 percent of its own budget, with the remainder coming from bilateral aid members and loans from the international community. As one leading scholar on the STP noted in 2006, two years after the NRF was implemented:

> Liberal democracy has not resulted in a sound economic policy, a more efficient state administration and a flourishing market economy... On the contrary, despite relatively large amounts of foreign aid and increasing donor activity in the country, there has been insufficient growth, while mass poverty and external debts have increased since 1991... Political behavior and institutional habits marked by political client elitism, nepotism, endemic corruption, the lack of accountability and transparency, and the weakness of the decision-making bodies have contributed considerably to this failure.[65]

The IMF partially agrees, and has cautioned that STP has a "narrow production and export base" which makes it "highly vulnerable to external shocks and dependent on aid inflows" and, despite debt relief, high rates of debt.[66]

This finding – that NRFs are insufficient by themselves to improve entire economies – is somewhat unsurprising given recent scholarship. One study from the Revenue Watch Institute looked at NRF performance across a variety of countries and concluded that "existing natural resource funds have often been inadequate in the effort to address the structural challenges of resource rich economies."[67] Humphreys and Sandbu write that:

> It is difficult to detect a consistent improvement of fiscal policy in countries with NRFs relative to those without them... A NRF is not a panacea for these problems and the incentives to spend too rapidly persist whether or not an NRF is established. The obvious difficulty is that NRFs are least needed when institutions are strong; but they are least likely to work in precisely those institutionally weak environments where they appear to be most needed. A fund that does not address the incentives of governments is subject to being abolished or ignored.[68]

Economist Ugo Fasano assessed six separate NRFs and noted that their outcome was "mixed" and reflective of "the challenges in adhering to the operational rules."[69] Another international comparative evaluation from the IMF concluded that there was little evidence to support the contention that adopting NRFs contributed to the soundness of a country's economic policies.[70] The lesson here may be that the ORML represents a success for STP but cannot be replicated effortlessly in other countries.

Opacity and corruption

There are, finally, signs that the ORML has failed to instill a complete culture of transparency within the government. There has been no oil production, only production signing bonuses, and therefore paltry revenues. One 2009 evaluation of STP's oil revenues noted that the government has still signed confidential agreements with oil companies and given them noncompetitive rights to blocks within the EEZ, even though requirements of the ORML prohibit it.[71] As this study concluded, "there is significant public distrust and lack of publicly accessible information on the current status of these agreements and efforts to renegotiate or revoke them has not been made public."[72] Corruption still afflicts STP as well, with Transparency International's corruption perception index ranking STP 111 out of 180 countries. As a partial explanation, that same year the minimum monthly wage for civil servants was only $46, creating a strong incentive for government employees to look for "supplemental" sources of income, much as they do in other countries such as Afghanistan, Somalia, India, Pakistan, and Haiti, to say nothing of Angola, Congo, Gabon, Cameroon, and Nigeria.[73]

Conclusion: lessons and implications

STP's Oil Revenue Management Law brings three conclusions to light. First, it supports the convention that NRFs can contribute to government savings and macroeconomic security. Oil proceeds, for example, have already generated millions of dollars in government revenue (through exploration royalties and signature bonuses) even though oil production has yet to commence, and are projected to increase national wealth in a Permanent Oil Fund 46 times the country's GDP, according to IMF projections, retaining more than $3 billion as a trust for future generations (as soon as somebody drills for oil, finds it, and starts production). The NRF has enabled the STP economy to begin to diversify activities away from get-rich, money-making schemes such as commemorative stamps and sex-related telephone calling centers. Although the NRF cannot take all of the credit, its implementation has coincided with a sevenfold increase in per capita GDP from 2003 to 2012, with inflation dropping, with the government reaching a balanced budget, and with numerous economic indicators and performance on transparency and governance indices steadily improving.

Second, STP's ORML showcases the importance of a President and a government having the foresight to ask for help when they need it. STP's prudency measures arose after President Menezes reached out to the economist Jeffrey Sachs and experts at institutions including Columbia University, the Open Society Institute, the World Bank, and the IMF. This enabled the government to build its capacity in issues relating to revenue management, corruption, transparency, conflicts of interest, civil society, and monitoring.[74] Apart from benefitting the people of STP, such actions have enabled the government to minimize the exploitation inherent in earlier petroleum contracts and production-sharing agreements with multinational oil companies. Without strong leaders on both sides – from the STP government, and from Sachs and his team – it is probable the ORML never would have occurred.

Third, however, is that, despite their utility, NRFs such as the ORML are not a cure-all. They only work when they combine a dedication to governance with sound fiscal and management policies and prudent oversight by independent bodies and civil society groups. As Professor Al Faruque recently put it, "while establishing a natural resource fund is an important step forward for transparency in revenue management, the success of the fund depends on whether it is able to combine accountability in its internal operation with sound economic policies. Such funds have been more successful in countries with a strong commitment to fiscal discipline and sound macroeconomic management."[75] Like many such tools, NRFs only work when their masters utilize them properly.

7
Intergenerational Equity and Solar Energy in Bangladesh

Introduction

The principle of intergenerational equity supposes that people have a right to fairly access energy services. Sadly, the current global energy system does not distribute its energy services equitably, and in 2009 approximately 1.4 billion people lived without electricity, 2.7 billion depended on wood, charcoal, and dung for domestic energy needs, and a further one billion people had access to electricity networks that was unreliable or unaffordable. *Soura Shakti* (SS) offers a blueprint for how planners around the world can rapidly expand access to modern energy services through the use of solar home systems (SHS). SS is the homegrown SHS program of Bangladesh's Infrastructure Development Company Limited (IDCOL). The program has, as of April 2012, successfully distributed 1.42 million SHS units throughout Bangladesh and is on target to reach four million by the end of 2015. To put this number in perspective, despite being one of the poorest countries in the world, Bangladesh has nine times the residential solar capacity of the United States, and it is home to the fourth largest market for photovoltaic (PV) installations after Germany, Japan, and Spain. In 2009, more than 99 percent of all SHS installations in Bangladesh occurred as part of the SS program. Even more impressive, the program achieved its targets three years ahead of schedule, without a single customer defaulting on a loan, and at $2 million below cost.

Intergenerational equity as an energy justice concern

Though many scholars have created strong grounds for the salience of intergenerational equity as a justice concern, some of the most compelling justifications come from John Rawls, Amartya Sen, and Martha C. Nussbaum.

The late Harvard professor John Rawls is perhaps the best-known advocate of a "contractionist" theory of justice, one based on the idea of a social contract. This concept is an intellectual response to Thomas Hobbes' famous notion that, without social order, society would devolve into anarchy and lives that were "solitary, poor, nasty, brutish, and short." Rawls' idea is that people make a contract with one another and agree to give up the private use of force and other social ills, such as thievery or murder, in exchange for peace, security, and mutual advantage; a society in which all people surrender some power before the law and due process.[1] As one study noted, "contract-theory states that only free and rational agreements that are reached in social cooperation can generate moral norms."[2] The idea shows that if we divest human beings from the artificial advantages they hold – wealth, rank, social class, education, musical ability, athleticism – they will agree to a contract of a certain sort.[3] Or, as Rawls himself writes, we should take as a starting point that "reasonable principles of justice are those everyone would accept and agree to from a fair position."[4] When applied to the realm of energy justice, this means thinking how we or a society would design a global energy system if we didn't know where we would fit within it – if we couldn't guarantee we would be the ones driving SUVs in the United States, or the ones collecting firewood in the rain in Sub-Saharan Africa. Rawls would argue that, under such a social contract, we would want an energy system fair and open to all, one that gives everybody an equal shot of receiving the energy services they need.

The Nobel-prize winning economist Amartya Kumar Sen expanded some of the ideas iterated by Rawls with his notion of "capabilities" for securing substantive human freedoms. The focus for Sen lies on what people can achieve with their lives, something he calls "functionings," or "the various things a person may value doing," such as being well fed, or receiving proper medical care, or being loved and safe, almost akin to states of mind, "doings," or "beings." Accordingly, functions go beyond income and economic goods; what counts is what people put those goods and services towards achieving. His notion of "capabilities" refers to the substantial freedoms people ought to have to enjoy the various combinations of functionings that they can realize. To the degree that poverty, or poor environmental conditions, or physical disabilities impede a person from experiencing the capacity or freedom to achieve the same level of functionings as others, then an injustice results.[5] For Sen, justice involves the removal of various types of un-freedoms that leave people with little choice or opportunity of exercising their reasoned agency.[6]

Martha C. Nussbaum, Distinguished Professor of Law and Ethics at the University of Chicago and a colleague of Sen, extended and built upon the notion of capabilities in her own work on justice. Her starting point is the fact that most ways of calculating the quality of life tend to distort human experience for the "poor" of the developing world because they assume living standards always improve as GDP does. This "crude approach" encourages many planners to work for economic growth without attending to the actual living conditions of poorer citizens. She develops what she calls the "Human Development" or "Capabilities" approach, which asks: "What are people actually able to do and to be?" and "What real opportunities are available to them?"[7] She uses the term "capabilities" to "emphasize that the most important elements of a people's quality of life are plural and qualitatively distinct: health, bodily integrity, education, and other aspects of individual lives cannot be reduced to a single metric without distortion."[8] The core of her theory of justice is a more concrete elaboration of the key capabilities people need to achieve their functionings:

1. Life.
2. Bodily health.
3. Bodily integrity.
4. Senses, imagination and thought.
5. Emotions.
6. Practical reason.
7. Affiliation.
8. (A relationship with) Other species.
9. Play.
10. Political and material control over one's environment, including political participation, freedom of speech, and the ability to hold property on equal basis with others.

Nussbaum does admit that "justice" does not provide any concrete guarantee that all people remain healthy or that they will have satisfying lives over most of their adulthood, but, rather, that all persons have the capability to realize these valuable functionings if they behave properly. Moreover, according to Nussbaum, a person lacking *any* of the ten capabilities fails to lead a fully human and dignified life. Similarly, deficiencies in any one capability do not offset or enhance the provision of others.[9] As she puts it, "the capabilities approach is fully universal: the capabilities in question are held to be important for each and every citizen, in each and every nation, and each person is to be treated as an end."[10] Energy justice, then, involves creating a life for households where they have the warmth, light, and cooked food to maximize their potential.

The global energy system, however, thwarts the ability of billions of people to do just that. For millions today, as in Biblical times, life without an energy grid means being "cursed" as "slaves" that do little more than "heave wood and haul water."[11] Some of us consume staggering amounts of liquid fuels and electricity – and have significantly large carbon footprints – while others go completely without them and contribute almost nothing to climate change. Some of us see energy as intertwined with our ideals of comfort, cleanliness, convenience, and luxury; others as a source of insecurity and drudgery.

The most common way of conceptualizing energy poverty relates to two metrics: (1) lack of access to electricity and (2) dependence on solid biomass fuels such as wood or dung for cooking and heating. According to the most recent data available, approximately 1.4 billion people still live without electricity, and an additional 2.7 billion people depend entirely on wood, charcoal, and dung for their domestic energy needs – figures reflected in Table 7.1.[12] Of about seven billion people in the world:

- 500 million consume an average of 10,000 kWh/household;
- 2.5 billion average 5,000 kWh/household;
- 2.5 billion average 1,000 kWh/household;
- 1.4 billion live without electricity at all.

Table 7.1   Number and share of population without access to modern energy services, 2009

		Without access to electricity		Dependence on traditional solid fuels for cooking	
		Population (million)	Share of total population (%)	Population (million)	Share of total population (%)
Africa		587	58	657	65
	Nigeria	76	49	104	67
	Ethiopia	69	83	77	93
	Congo	59	89	62	94
	Tanzania	38	86	41	94
	Kenya	33	84	33	83
	Other Sub-Saharan Africa	310	68	335	74
	North Africa	2	1	4	3
Asia		675	19	1,921	54
	India	289	25	836	72
	Bangladesh	96	59	143	88
	Indonesia	82	36	124	54
	Pakistan	64	38	122	72
	Myanmar	44	87	48	95
	Rest of developing Asia	102	6	648	36
Latin America		31	7	85	19
Middle East		21	11	0	0
Developing Countries		1,314	25	2,662	51
World		1,417	19	2,662	39

Source: International Energy Agency.

This energy deprivation for the bottom of the pyramid results in four major and interrelated sets of negative consequences: poverty, death, gender inequality, and environmental degradation.

One of the more persuasive reasons energy poverty is a justice concern, like fuel poverty in Chapter Three, relates to equity – it's unfair that the poor must spend more of their income on energy. One survey found that roughly 33 percent of households in the poorest income quintile suffered from energy poverty compared with fewer than one percent of those in the richest quintile.[13] Another study of energy use in Latin America and the Caribbean found that the highest quintiles of households – the so-called "richest" – consumed three to 21 times more energy than the lowest quintiles, the so-called "poor." This means that access to energy, or lack of it, can each reflect and worsen social inequality.[14] The study found, for example, that, in all countries analyzed, the poor consumed less energy than other strata of households, but spent more of their income on it. The price per unit of energy was usually higher, since poor households have difficulty accessing electricity grids or liquid fuels which have higher energy densities. Rural households tend to be poorer and consume much less energy than urban households, and in rural areas fuelwood, the most common source of energy, is usually harvested in unsustainable ways, with severe impacts on forest health and the health of those using it, as discussed below.

However, the problem of energy equity and access is not confined to rural areas and poor countries alone. In terms of urban areas, one recent investigation of 34 cities in Bolivia, Botswana, Burkina Faso, Cape Verde, Haiti, India, Indonesia, Mauritania, Philippines, Thailand, Yemen, Zambia, and Zimbabwe concluded that "the poor in urban areas of developing countries face special problems in meeting their basic energy needs." It noted that most urban poor are migrants and continue to rely on traditional fuels they collect on the periphery of urban areas, and also that they pay higher prices for usable energy because of the inefficiency of stoves and lamps.[15] Consider one comprehensive analysis which looked at middle-income consumption patterns in 14 countries and found extreme asymmetries in wealth. In India, the upper-middle-income households account for less than one-eighth of the population, but have two-fifths of the country's purchasing power and account for 85 percent of private spending. The study noted that the per capita energy consumption from these more affluent homes is 15 times greater than the rest of India's population.[16]

In terms of wealthier countries, this disparity between the energy-rich and poor can be expressed through a Gini coefficient or Lorenz curve, a tool that describes the degree of income concentration related to energy consumption (varying between zero for perfect equality and one for maximum inequality). Energy inequality seems to plague rich and poor countries alike; it is not a problem relegated to only the developing world. One study appraised the equity of energy use in El Salvador, Kenya, Norway, Thailand, and the United

States.[17] Even the most equitable country, Norway, saw half of residential electricity being used by only 38 percent of customers; in the United States half of household electricity was used by 25 percent of households; in El Salvador, 15 percent; Thailand, 13 percent; Kenya, six percent. Other studies have confirmed similar trends at the national level in Greece,[18] Mexico,[19] and the United Kingdom,[20] as well as at the village and household level in India[21] and at the technological level (for things like clothes washers, clothes driers, refrigerators, and freezers).[22] Another study looked at fossil fuel consumption in the United Kingdom from 1968 to 2000 and found that higher incomes result in greater inequality related to the consumption of fuel and light, car use, recreation, and international travel.[23] Yet another found a clear relationship between income and car ownership, with the richest decile 11 times more likely to have the use of a private car than households in the poorest decile, of which fewer than one in ten own their own automobile.[24] In Australia, middle and upper-income households consumed as much as four times more fuel, light, power and transport services than lower-income homes.[25] In this way, energy poverty is an issue of distributional injustice, linked to "wider established concerns over inequality, poverty and the interests of the least advantaged."[26]

Regardless of where they occur, poverty and energy deprivation go hand-in-hand, with energy expenses accounting for a significant proportion of household incomes in many developing countries. Generally, only three to four percent of household income goes towards energy in middle and upper-income households. In contrast, 20 to 30 percent of annual income in poor households is directly expended on energy fuels, and an additional 20 to 40 percent is expended on indirect costs associated with collecting and using that energy, such as health care expenses, injury, or loss of time. In other words, the poor bear on average eight times more of a burden for the same unit of energy as other income groups.[27] In extreme cases, some of the poorest households directly spend 80 percent of their income obtaining cooking fuels.[28] In Nepal, I visited a woman who proudly displayed a single fluorescent light bulb – which cost less than $3 – as her "most prized possession in the world."

More seriously, energy poverty literally kills people. It has grave and growing public health concerns related to indoor air pollution (IAP), physical injury during fuelwood collection, and lack of refrigeration and medical care. By far the most severe of these is IAP. Most homes without access to modern forms of energy cook and combust fuels directly inside their home. Burning firewood, dung, and charcoal is physiologically damaging, akin to living within a giant smoking cigarette. Almost three-quarters of people living in rural areas and half (45 percent) of the entire global population rely on wood and solid fuels for cooking.[29] Yet, as the WHO explains:

> The inefficient burning of solid fuels on an open fire or traditional stove indoors creates a dangerous cocktail of not only hundreds of pollutants,

primarily carbon monoxide and small particles, but also nitrogen oxides, benzene, butadiene, formaldehyde, polyaromatic hydrocarbons and many other health-damaging chemicals.[30]

There are damaging spatial and temporal dimensions to such pollution. Spatially, it is concentrated in small rooms and kitchens rather than outdoors, meaning that many homes have exposure levels to harmful pollutants 60 times the rate acceptable outdoors in city centers in North America and Europe.[31] Temporally, this pollution from stoves is released at precisely the same times when people are present cooking, eating, or sleeping, with women typically spending three to seven hours a day in the kitchen.[32] Even when these homes have a chimney and a cleaner-burning stove (and most do not), such combustion can result in acute respiratory infections, tuberculosis, chronic respiratory diseases, lung cancer, cardiovascular disease, asthma, low birth weights, diseases of the eye, and adverse pregnancy outcomes; as well as outdoor pollution in dense urban slums that can make air unbreathable and water undrinkable.[33]

Strikingly, IAP ranks *third* on the global burden of disease risk factors at four percent, coming after only malnutrition (16 percent) and poor water and sanitation (seven percent).[34] This places it ahead of physical inactivity and obesity, drug use, tobacco use, alcohol use, and unsafe sex. IAP is currently responsible for a shocking 1.6 *million* deaths each year – more than 4,000 deaths per day, or almost three deaths per minute. The cost of this burden to national health care systems, not reflected in the price of energy, is a

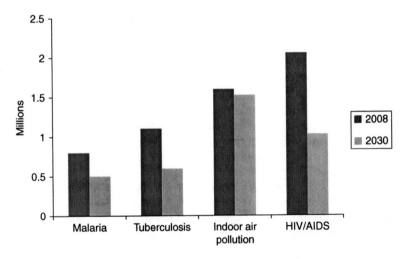

Figure 7.1   Annual deaths worldwide by cause, 2008 and 2030
Source: Adapted from World Health Organization.

whopping $212 billion to $1.1 trillion.[35] Almost all of these deaths occur in developing countries, and more than half occur in children.[36] Put in perspective, the number of deaths from IAP is already greater than those from malaria and tuberculosis.[37] More worryingly, Figure 7.1 shows that by 2030 the number of deaths from IAP will likely be greater than from malaria, tuberculosis, and HIV/AIDS.[38]

The IAP statistics for some particular countries are frightening. In Gambia, girls under the age of five carried by their mothers while cooking have a six times greater risk of lung cancer than if their parents smoked cigarettes and were not exposed to IAP from cooking.[39] From 2005 to 2030, ten million women and children in Sub-Saharan Africa will die from the smoke produced by cooking stoves.[40] One investigation of four provinces in China found that IAP affected every person in a rural home, and also that inefficient combustion of fuel is not the only problem; so is food drying and storage.[41]

Unfortunately, IAP is not the only health consequence of energy poverty. Women and children face exposure to health-related risks during the burdensome and time-intensive process of collecting fuel. Common injuries include back and foot damage, wounds, cuts, sexual assaults, and exposure to extreme weather. The large number of daily hours women need to collect and use solid fuel leaves them with no other option than to take young children with them, in essence exposing both to the same health impacts.[42] Ten thousand fuelwood carriers in Addis Ababa, Ethiopia, supply one-third of the wood consumed by the city and suffer frequent falls, bone fractures, eye problems, headaches, anemia, internal body disorders, and miscarriages from carrying loads often equal to their body weights. Fuel collection also places women in areas of physical or psychological violence. Hundreds of documented cases revealed Somali women being raped while collecting fuel, and women in Sarajevo faced sniper fire to collect biomass.[43] In India, for instance, the typical woman spends 40 hours collecting fuel per month during 15 separate trips, many walking more than a 6-kilometer round trip.[44] This amounts to 30 billion hours spent annually (82 million hours per day) collecting fuelwood, with an economic burden (including time invested and illnesses) of $6.7 billion (300 billion rupees) per year.[45]

Countries without access to modern energy also tend to have more dilapidated health systems. Consider that, compared with developing countries, infant mortality rates are more than five times higher in energy-poor countries, as is the proportion of children below the age of five who are malnourished (eight times higher), the maternal mortality rate (14 times higher), and proportion of births not attended by trained health personnel (37 times higher).[46] In Papua New Guinea, for example, many doctors in rural areas without electricity perform births while holding a flaming torch in one hand for light, operating with the other, subjecting pregnant women to severe burns. In Sri Lanka, midwives and those assisting women with giving birth sometimes knock over kerosene lanterns, resulting in "thousands of serious burns" every year.

Furthermore, indirect health effects occur when traditional fuel becomes scarce or prices rise. Meals rich in protein, such as beans or meat, are avoided or undercooked to conserve energy, forcing families to depend on low-protein soft foods such as grains and greens, which can be prepared quickly. In other cases, families stop boiling drinking water when faced with an energy shortage.[47] Unboiled and unclean water then contributes to some of the 4,500 deaths throughout the world every day from cholera, typhus, and other diseases.[48]

However, small-scale renewable energy technologies – SHS, residential wind turbines, biogas digesters and gasifiers, microhydro dams, and improved cookstoves – offer households and communities the ability to tackle extreme poverty and raise standards of living.[49] Put another way, they enable us to meet our international equity obligations.

Case study: IDCOL's Solar Home Systems Program

Bangladesh is one of the world's most densely populated and poorest countries. About four-fifths of its rapidly growing population of 144 million inhabitants live in rural areas, 40 percent of the entire population live below the poverty line, and one-fifth of all Bangladeshis persist in what the World Bank calls "extreme poverty."[50] Many of the country's low-lying areas are prone to perennial flooding, erosion, and saltwater intrusion of water supplies. Per capita GDP is an extremely low $772 (adjusted for purchasing power parity) and about one-quarter of young adults aged 15 to 24 are still illiterate.[51]

Bangladesh is also one of the world's lowest producers and users of energy, with total primary energy supply at only about 150 liters of oil per capita per year – an average *less* than that for the entire continent of Africa.[52] Almost four decades after the country's independence, shortages of natural gas, which provides 70 percent of commercial energy (electricity and liquid fuels), are now common. The most recent numbers offered by the Rural Electrification Board (REB) show an estimated 71 percent of households without access to the electricity grid, numbers presented in Table 7.2.[53] Indeed, Table 7.3 shows that, due to natural disasters, some years, such as 2008 to 2009, actually saw a *negative* electrification rate. Moreover, no single political party has been elected for two terms in a row for the past two decades, creating a lack of continuity in energy policy planning and implementation.

As a result, traditional forms of wood, twigs, agricultural residues, and other forms of biomass meet an enormous 65 percent of all energy needs in the country. The United Nations Development Program calculates that kerosene, natural gas, and electricity combined meet less than one percent of rural household energy needs.[54] For comparison, energy consumption in

Table 7.2 Electrification of villages and households in Bangladesh, 2010

Division	% Villages Grid Connected	% Households Grid Connected
Chittagong	78	39
Dhaka	80	44
Khulna	50	11
Rajshahi	53	20
All	66	29

Source: Adapted from World Bank Group.

Table 7.3 Rates of grid electrification in Bangladesh, 2003–2010

Year	2003–04	2004–05	2005–06	2006–07	2007–08	2008–09	2009–10
Villages Electrified	41,125	44,224	45,794	46,720	48,327	47,641	48,682
% of increase	7.06	7.54	3.55	2.02	3.44	−1.42	2.19
Number Added	2,711	3,099	1,570	926	1,607	−686	1,041

Source: Adapted from World Bank Group.

Bangladesh is less than one-tenth the global average, and 96 million people lack access to the grid.[55] National planners anticipate a need for $10 billion worth of investment by 2020 only to maintain the grid and stop load shedding. Given the current rate of electrification, rural Bangladeshis will have to wait 40 years for likely grid coverage, a feat complicated due to the country's many rivers and waterways, which make transmission and distribution of electricity difficult.[56] Moreover, considering the power sector's high dependency on increasingly scarce natural gas and Bangladesh's vulnerability to natural disasters, there is an urgent need for the country to diversify its sources of electricity.[57]

Such socioeconomic, climatic, and political vulnerabilities make it all the more impressive that the Infrastructure Development Limited Company (IDCOL), a government-owned investment company, has financed 1.42 million SHSs to rural users and is set to finance four million units by the end of 2015. To put this number in perspective, in 2008 Bangladesh had almost *nine times* the number of installed residential solar systems as the United States,[58] and the fourth largest market for PV installations, after Germany, Japan, and Spain. Even more salient is the fact that the program has achieved such rapid commercialization in less than ten years, ahead of schedule, and below estimated costs. Former US presidential candidate John

Kerry, also chair of the Senate Foreign Relations Committee, has called the IDCOL program a "literally life-altering project" for the communities it has reached.[59]

History

IDCOL began its SHS program, or *Soura Shakti*, which means "solar energy" in Bengali, in 2003 with assistance from the World Bank and Global Environment Facility, and set a target of installing 50,000 SHS by 2008. In a departure from previous "grant-based" approaches, IDCOL was asked to focus on "developing a consumer-focused integrated sustainable model" that would "encourage customers to install a system because it was fulfilling their needs and not because of any subsidy."

IDCOL itself is an entity created by the Bangladeshi government in 1997 to finance infrastructure projects. It receives most of its funding from multilateral development institutions such as the World Bank, Asian Development Bank, Islamic Development Bank, Deutsche Gesellschaft für Internationale Zusammenarbeit (GIZ), KfW Bankengruppe, Global Partnership on Output-Based Aid (GPOBA) and SNV Netherlands Development Organization. IDCOL's largest sector is energy and power, and renewable energy (SHS, biogas, and biomass) represents 64 percent of its investment in energy, or about 56 percent of its entire $361 million investment portfolio.

IDCOL is the largest financial institution in the country promoting renewable energy and is well situated to promote community-based projects in off-grid areas. One advantage the institution has is that it is exempt from provisions by central banks on capitalization of financing, which require commercial banks to operate with certain debt to equity ratios and short payback periods. IDCOL, in contrast to these banks, is permitted to operate with longer paybacks and lower interest rates because the government recognizes that different models are needed to finance development and infrastructure projects. IDCOL was designed to serve as a catalyst to fill the gap in development financing as opposed to rapidly making money like commercial banks.

For those who have never seen one, the typical SHS consists of a solar photovoltaic module, battery, charge controller, and lamp, shown in Figure 7.2. Major advantages of these units compared with other sources of energy in Bangladesh include ease of assembly and operation, the ability for customers to generate their own energy without bills (other than those related to the capital cost of the system), and their durability and long lifetime, often in excess of 20 years.[60] IDCOL customers in off-grid and rural areas could, when the program started in 2003, choose from a variety of technologies from a 20 Watt-peak (Wp) system to a 130 Wp system presented in Table 4 (though the inventory of systems was later expanded to include 10 Wp systems, to better accommodate the poorest customers).

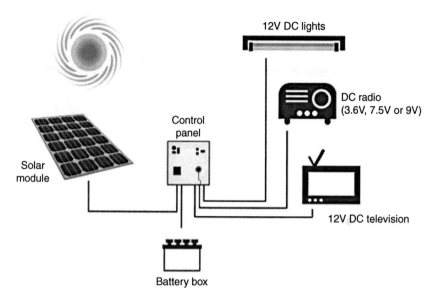

Figure 7.2 A typical Solar Home System (SHS) in Bangladesh

Table 7.4 Details of eligible Solar Home Systems under IDCOL's SS Program, 2003

Capacity	Total Load	Operating Hours	Cost (in USD)
20Wp	Lamp: 2 (5 W each) Mobile Charger: 1	4–5 hours	170
50Wp	Lamp: 4 (7 W each) Black and White TV: 1 Mobile Charger: 1	4–5 hours	380
85Wp	Lamp: 9 (7 W each) Black and White TV: 1 Mobile Charger: 1	5–7 hours	580
130Wp	Lamp: 11 (7W each) Black and White TV: 1 Mobile Charger: 1	7–8 hours	940

Source: Adapted from Infrastructure Development Company Limited.

Part of the reason these systems work so well throughout the country is because Bangladesh receives an abundant amount of radiation (about 4 to 6.5 kWh per square meter). Moreover, Bangladesh has a history of SHS use going back to the 1970s, when the REB started promoting it in rural areas as a substitute for grid extension. SHS can be easily installed and maintained with minimal amounts of training, have lower operating costs than other renewable energy options, and do not produce unpleasant smells or pollution during operation (unlike biogas units).[61] As one high-ranking government official told the author:

> We don't even know the technical potential for small-scale wind energy in Bangladesh, no one has done the assessments or wind mapping, the country has insignificant fossil fuels other than natural gas, which is running out, and we sit on a really flat delta, meaning we have lots of water but it's not moving, making hydroelectricity difficult. Our susceptibility to natural disasters also creates a strong incentive to decentralize and diversify sources of energy production. Demarcation and political issues prevent us from importing a large amount of electricity from West Bengal (India) or Myanmar. SHS are an obvious fit to the contextual challenges facing Bangladesh.[62]

These factors likely explain why SHS, though still limited in their overall diffusion, constitute more than 90 percent of current renewable energy deployment in the country.[63]

IDCOL established three committees to oversee and manage the SS program. A Participating Organization (PO) Selection Committee consisted of representatives from different organizations, including relevant government ministries, to select eligible POs that would implement the program on the basis of experience in micro-finance and institutional strength. POs are expected to make money on their SHS projects and expand to form a national distribution network.[64] The program initially had five POs but has now expanded to more than 30, including:

- Grameen Shakti
- BRAC Foundation
- Srizony Bangladesh
- COAST Trust
- Thengamara Mahila Shabuj Shangha
- Integrated Development Foundation
- Centre For Mass Education in Science
- Upokulio Bidyuatayon O Mohila Unnayan Shamity
- Shubashati
- Bangladesh Rural Integrated Development For Grub-Street Economy
- Padakhep Manbik Unnayan Kendra
- Palli Daridra Bimochan Foundation
- Hilful Fuzul Samaj Kalyan Sangstha
- Rural Services Foundation
- Network for Universal Services and Rural Advancement (NUSRA)

- DESHA
- Al-Falah Aam Unnayan Sangstha (AFAUS)
- Association for Village Advancement (AVA)
- DABI Moulik Unnayan Sangstha
- Ingen Technology Ltd.
- Rimso Foundation
- Rural Energy and Development Initiative (REDI)
- Green Housing & Energy Limited (GHEL)
- Shakti Foundation for Disadvantaged Women (SFDW)
- Association for Development Activity of Manifold Social Work (ADAMS)
- Resource Development Foundation (RDF)
- RISDA Bangladesh
- Bright Green Energy Foundation (BGEF)
- SolarEn Foundation (SEF)
- Patakuri Society

In tandem with the creation of a PO Selection Committee, a Technical Standards Committee was formed, consisting of technical experts from local universities and engineering departments of government bureaus. This Committee determined standards, reviewed product credentials, and approved eligible equipment. An Operations Committee consisted of program heads from all POs, and representatives from IDCOL manage and monitor what one company official called "all operational aspects of the program."

Under its financial model, IDCOL first recruits POs through a request for proposals and competitive bidding process. Once selected, POs sign a Participation Agreement, or PA, to distribute SHS, but IDCOL gives financial support only through a four-stage process. The first segment involves a capital buy down grant, or "Grant A," which enables POs to sell SHS at a slightly reduced price so that communities can afford them. The second segment involves an institutional development grant, or "Grant B," given to POs to extend credit to households, and also enabling POs to build their capacity by hiring staff and training employees in micro-finance and credit monitoring. Grants A and B are intended to permit POs to purchase the technology below market rates and also provide loans to customers, in essence lowering the price of SHS and increasing the institutional strength of distributors at the same time. To promote competition, expenditures are reduced in amount over time as more SHS capacity is installed, an element called "a phased reduction of grants," depicted in Table 7.5. The third segment, once a certain amount of capacity has been reached, involves a Technical Assistance grant (usually for a few hundred thousand dollars), which can be utilized by POs for advanced training and promotional campaigns, such as selling SHS via motorbikes, purchasing and using computers, and tweaking advertising

Table 7.5 Phased reduction of buy-down and Institutional Development Grants under the IDCOL SS, 2012

	Amount of Grant Available per SHS		
Item	Total	Buy-down grant	Institutional Development Grant
First 20,000 SHS	$90	$70	$20
Next 20,000 SHS	$70	$55	$15
Next 35,000 SHS	$50	$40	$10
Next 88,160 SHS	$48	$48	$10
Next 35,000 SHS	$46	$48	$8
Next 235,000 SHS	$44	$48	$5
Next 100,000 SHS	$36	$33	$4
Next 72,000 SHS	$30	$27	$3
Amended Grant in 2012	$25	$23	$2
Amended Grant in 2013	$0	$0	$0

Note that, starting in 2013, no grant will be provided for SHS larger than 30 Wp. Only smaller SHS will receive a grant at $20. This is to enable the poorer segment of the community to purchase SHS, as per their need.*Source*: Adapted from Infrastructure Development Company Limited.

efforts. The fourth segment is a low-interest refinancing loan, sponsored by the World Bank, Asian Development Bank (ADB), Islamic Development Bank (IDB), and Bangladeshi government, extended to the POs when they need extra resources to improve quality and operations. It enables refinancing 80 percent of the credit sale granted to households, or about $232 of every $290 in sales, over seven to ten years at a six to nine percent interest rate (including a one to two-year initial grace period).

Taken collectively, these elements combine to lower the cost of purchasing a SHS. A 50 Wp system that would normally cost $380 would require a household to pay only $51.24 for that system upfront, the rest formulated into a low-interest loan paid back over two to five years with monthly payments of about $10.

In the words of one of the program's managers, IDCOL attempted to "design the whole thing to learn from past failures" and "rigorously matched specific program elements to overcome identified challenges."[65] They wanted to create a program where "all aspects of the SHS supply chain could be concentrated in one place, from supply and financing to installation, maintenance, and tariff collection." They sought to "permanently deploy outlets in villages that would be financially self-sustainable." And they "matched specific program components to a series of barriers that previous household surveys and studies had identified," summarized in Table 7.6.[66]

Four things stand out to make the program notable: a focus on commercial viability, institutional diversity, quality technology, and distributing

Table 7.6   Identified challenges and matched IDCOL Program components

Identified Challenges	IDCOL Solar Program Components
Lack of capacity of the government, private sector and financial sector	A private–public partnership model was selected to encourage micro-financing to make units more affordable
Lack of tailored financing package	A Capital Buy-down Grant was created to reduce the system price along with an Institutional Development Grant and long-term refinancing channeled to POs
High initial cost of solar equipment	Capital Buy-down Grant reduces system price and systems are also sold on credit to households to ensure affordability
Adequate institutional capacity of the executing agencies	Institutional Development Grant and long-term refinancing are channeled to the executing agencies for capacity-building
Lack of an effective business model for private sector participants	A social enterprise model was created so that implementation occurred by a combination of micro-finance institutions and communities themselves
Lack of awareness among potential customers	Joint training, marketing and promotional activities are continually undertaken to increase awareness among potential customers
Absence of uniform product quality and lack of tools for ensuring quality	Technical Standard Committee must approve standard equipment and accessories used in the program; and POs provide after-sales service and stringent monitoring
Absence of any administrative agency to monitor quality of product and service	Three-tiered monitoring system in place consisting of IDCOL Inspectors, reinspection by regional supervisors, and donor inspection by POs
Absence of local related and support industries	Program has supported a domestic manufacturing base for SHS components such as batteries and charge controllers and also solar photovoltaic assembly plants in Bangladesh

Source: Adapted from Infrastructure Development Company Limited.

risk. Rather than previous approaches, which were largely grant-based, perhaps the most distinct element of the SS program is its focus on *commercial viability* and an innovative financing model, not only for households and end-users, but also for development organizations and partners. As one interview respondent from the private sector put it, "The intention is to get rid of the grant component altogether at some point. If a nation is hooked on charity, they will only want more charity. The survivor attitude will disappear."[67] Thus, the central idea was "to make SHS projects commercially viable, and only support partners and institutions that could run like a business; if they would not turn a profit, we would not do a project."[68] After the capital buy-down grant, a client or user must still finance a minimum ten percent of the cost of a system through an upfront down-payment. POs, too, must demonstrate that they are collecting a majority of payments on time or they are suspended from the program. As one official from IDCOL stated, "we are serious about the performance of POs, if any particular branch or village office falls below 50 percent collection efficiency, we stop the financing and suspend them temporarily until they reach efficiency again."[69]

Second, the program emphasizes *institutional diversity* by involving a network of five types of collaborators that all have their "skin in the game." IDCOL covers project finance and management, multinational and national donor agencies provide additional funding, nongovernmental organizations and micro-finance institutions operate as POs, manufacturers and suppliers interact with the IDCOL Technical Standards Committee, and energy development professionals assist with expansion and rural development. The "most important link in this chain" is the POs, or, as one respondent from the government put it, "getting good quality implementing agencies is the hardest part, especially here in Bangladesh, which is why every PO is chosen carefully through peer review and competitive bidding."[70]

Third, the program focuses on *improving technology*. One respondent from a PO in the program argued that "Bangladesh has no problem with fake or substandard panels or equipment because suppliers are always following specifications; every panel has a serial number and a warranty, and all SHS are checked by a technician to be fully operational before customers have to make payments."[71] Another, from a solar manufacturer, mentioned that "IDCOL does strict internal auditing to ensure quality standards are met by POs and all manufacturers."[72] Yet another, from civil society, remarked that "we have a big team of about 50 people doing quality control and standardization, traveling every day to check on systems installed who verify about 50 percent of all installations in a given area, and who also have the power to penalize unsatisfactory performance."[73] The Technical Standards Committee and these teams, headed by professors and engineering experts, not only improved the quality of panels, but also "set standards for inverters, charge controllers, batteries, and installation procedures, along with adhesives, mounting, and maintenance."[74]

Fourth, the program *distributes risk* for POs and SHS users in a variety of ways. Households must pay a percentage of the upfront cost of a SHS; the remaining amount is split between IDCOL and the particular PO. So, if a SHS were to cost $100, a customer would have to pay $10 upfront and then the remaining $90 would be split so that IDCOL takes responsibility for $72 and the PO takes responsibility for $18. As one respondent from a PO put it, "the program nicely distributes risk to all three key players: households have to put up a share, the PO has an investment, and IDCOL is at risk as well, though this risk is minimized since the POs are responsible for collecting payments directly from consumers. The 10 percent risk to households may not sound like much, but it's up to three months of household income, so it is taken seriously."[75] Also, once a household is connected to the electricity grid, it can resell the SHS back to the POs at cost so that customers don't "lose out" by grid electrification.

By almost any measure, IDCOL's SS has been a great success. The initial target was to finance 50,000 SHS by June 2008, achieved in September 2005 – three years ahead of schedule and at $2 million below projected costs, due mostly to improvements in technology and declining installation fees. As Figure 7.3 shows, at the end of 2009 IDCOL had installed 420,000 SHS and reached 800,000 in 2010 – meaning systems benefitted about 3.4 million people – and, if all goes well, will reach four million households by the end of 2015. Table 7.7 also shows that more than 1.2 million units have been installed as of April 2012. IDCOL has financed $104 million under the program but not a single customer has defaulted

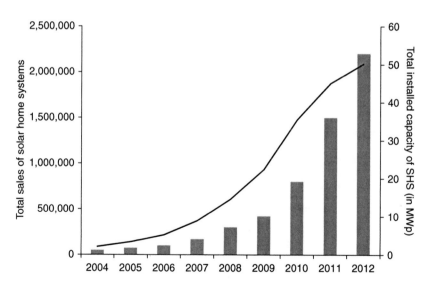

Figure 7.3   Solar Home Systems and total installed capacity achieved under the IDCOL Program, 2004 to 2012

Note that the bar graph and legend on the left show total sales of SHS; the line and legend on the right show total installed capacity.

Source: Adapted from Infrastructure Development Company Limited.

Table 7.7   Partner organization installation of SHS as of April 2012

Partner Organization	Number of SHSs Installed
Grameen Shakti	795,957
RSF	216,434
BRAC	77,019
Srizony Bangladesh	58,927
Hilful Fuzul Samaj Kalyan Sangstha	37,078
UBOMUS	25,234
BRIDGE	20,449
Integrated Development Foundation	14,238
TMSS	13,059
PDBF	10,672
SEF	21,720
AVA	12,817
DESHA	10,931
BGEF	16,995
RDF	20,027
COAST	6,181
INGEN	9,871
CMES	5,714
NUSRA	9,372
RIMSO	8,196
Shubashati	5,370
REDI	5,711
GHEL	6,138
SFDW	9,485
PMUK	2,166
Patakuri	3,409
ADAMS	2,848
AFAUS	1,161
RISDA	1,552
Xenergeia	320
Other	389
Total	1,429,440

Source: Adapted from Infrastructure Development Company Limited.

on a loan, and not a single PO has fallen below 50 percent collection efficiency. IDCOL reported a return on equity of 25.5 percent in 2008 and 28.2 percent in 2009. During that same year, more than 99 percent of all SHS installations in Bangladesh occurred as part of the program. Following this success, the World Bank, Asian Development, Islamic Development Bank, and other actors have extended their support and scaled up financing, and officials from Ethiopia, Ghana, India, Nigeria, and Tanzania have approached IDCOL about replicating the program in their own countries.

Benefits

The IDCOL program has five distinct benefits: (1) employment, (2) reliable off-grid electrification, (3) affordability, (4) income generation, (5) better technology.

Employment

Thanks to SS, employment related to POs and distributors of SHS equipment, as well as local jobs in maintenance and installations, has expanded considerably. As of 2011, more than 12,000 people had full-time jobs directly related to the program, whether installing SHS on household roofs and manufacturing components for SHS like batteries and inverters, or refurbishing load controllers, or designing and improving other SHS-related equipment. Benefits for rural communities have been called "significant" as women become trained in maintenance and installation.

Reliable off-grid electrification

Grid electricity is available predominately in and around the urban areas of Chittagong and Dhaka. An unreliable grid, as well as "stalled" plans to extend grid electrification to rural areas, makes SHS "the only real option for getting energy services to poor, rural populations." As one manager from a multinational institution elaborated:

> There is a motivation for SHS systems in both urban and rural areas. In urban areas, the power system is notoriously unreliable, it goes off and on all the time. Power generation is managed by the government, but they are inefficient and their equipment is aging. The Rural Electrification Board, responsible for grid extension, has no real plan to extend the grid anytime soon. This means for the near term, people can either live completely without access to energy services, or they can adopt SHS.[76]

Others called plans for rural electrification through grid extension "corrupt," "flawed," and "unlikely."[77] One respondent from an NGO even joked that, while the power companies say they are concerned with extending the grid, "they speak more about the issue at the same time they work less to achieve it."[78] Part of this is speed: it can take two years just to get a new power plant approved and another five to get it built, whereas SHS can be deployed in a matter of days.

Other respondents cautioned that the private sector was unlikely to fill the gaps left in grid coverage. As one from the government noted:

> Most banks and companies in the private sector are not interested in energy poverty, they are not concerned about development or infrastructure. There are some minor players, but the private banks like bigger chunks of money, more capital intensive projects. They are comfortable

with power plants or LNG facilities, but not interested beyond 5 to 7 years. Yet successful energy poverty programs need to run for 10 to 20 years. With major private players earning a 12 percent return on investments in five years, they see no incentive to earn 3 to 4 percent over ten to twenty years with poor communities.[79]

Affordable energy services

While SHS represent a significant capital investment for poor households, they tend to pay for themselves within the first five years. This is because most rural households meet their energy needs with kerosene, which can cost $0.70 per liter or as much as $25 per month – meaning a SHS will pay for itself in two to four years, producing net savings thereafter. Indeed, REB surveys continue to show kerosene usage to be the primary source for lighting across all districts in Bangladesh, including grid-electrified Chittagong and Dhaka, ostensibly underscoring the unreliability of those grids.[80]

One interesting dimension of this benefit is the targeted nature of the IDCOL program for "the poorest" households, that is, those living in "extreme poverty" on less than $1 per day. Access to energy in Bangladesh is highly uneven between income groups, with the lowest deciles using slightly more than 2 kilograms of oil equivalent (KGOE) per capita per month but wealthier deciles using almost three times as much.[81] To account for these inequalities of access, IDCOL started promoting cheaper, smaller 10 Wp systems to be explicitly targeted at the poorest households. Although some rural residents expressed a differing view below in our section on challenges, this has resulted in what one IDCOL manager called "a segmentation of the market" with "10Wp and 20 Wp systems selling predominately to the poorest households," "40 Wp and 50 Wp selling to the average rural family that needs enough electricity to run a black and white television, four lights, a mobile phone charger and possibly a small fan," and "130 Wp systems selling to what might be considered the lower middle class of Bangladesh with enough energy for fans, lights, a TV, mobile phone charger, and radio."[82] To ensure that wealthier families do not take undue advantage of the program, however, IDCOL limits the subsidy to one system per family, so no household can purchase multiple systems, and sets a ceiling on eligible capacity (130 Wp is the highest that is covered) and a limit on subsidized refinance (no more than $285 per system per household).

Income generation

As one respondent from a solar manufacturer noted, "the IDCOL program is supposed to be commercially viable but we don't try to control people's behavior or limit what they can use their electricity for, we don't tell them they have to purchase lights or whether they should have a television or not."[83] Because of this flexibility, some SHS households – we could not confirm how many – use their systems to promote a rich mosaic of income-generating activities. One family we met utilized their 50 Wp system to keep one light for themselves and rent out three lights to their neighbors, earning about $8 per month (or more than three-quarters of their monthly installment fee). Others utilized their SHS to charge mobile phones, making $0.14 per phone charged. Still others have started rural commercial enterprises on a household scale, by converting their home into a restaurant or building a SHS tea stand on their property, refrigerating vaccines for local health clinics, starting mini-cinema halls (think a TV, DVD player, and six plastic chairs), creating study areas for students at night, pumping water for farms (either their own, or for somebody else for a fee), sewing in the evening, or selling goods after dusk. As one respondent from IDCOL put it succinctly, "the biggest advantage of a SHS is its ability to generate income for families and communities, its ability to increase working hours, learning hours, productive hours, SHS is a revolution to these communities, a quiet revolution in their social and economic structure."

Indeed, these benefits have been confirmed by three other studies. The first, from the REB, surveyed more than 11,000 residential, commercial, and industrial energy users, including those with SHS and those without.[84] The survey found that the biggest difference between SHS users and non-users was the number of income earners per household, with non-SHS-electrified homes having 1.5 to 1.6 but SHS-electrified homes boasting 2.4 to 2.5. Following through with a qualitative study of six districts (a district is the equivalent of a state) – Barisal, Bagerhat, Moulavibazar, Manikganj, Rangpur, and Comilla – the study also found that before grid connection or reliance on SHS houses were dependent on kerosene lamps and hurricane lamps, which emitted soot and fumes indoors, causing pollution and health problems, constrained the ability for children to study and shops to stay open after dark, and limited the networking and income-generating abilities of women. These communities had limited use of mobile phones and did not receive timely information about current events and news. Electrified homes, by contrast, were cleaner and more prone to children studying and staying in school. Electricity enabled shops, stores, rice mills, and small traders to remain open for longer hours and women to feel more secure when traveling to hospitals, schools, and learning centers. Women became more involved in community activities such as sewing at night and running shops, and were able to rely on mobile phones and televisions to become more aware about issues relating to reproductive health, children's health, family planning, early marriage, dowries, and forestry. Many were even able to talk directly to doctors during medical emergencies. Indeed, the survey found that SHS produced a variety of services well beyond lighting, to include cooking with electric rice boilers, television, and radios (see Table 7.8).[85]

Table 7.8   Use of solar electricity by Bangladeshi households

	Barisal	Chittagong	Dhaka	Khulna	Rajshahi	Sylhet	Total
Lighting	73.6%	76.3%	80.3%	84.9%	78.1%	81.0%	78.0
Cooking	5.6%	3.1%	0%	1.9%	0%	0%	2.3
Both	1.4%	5.6%	1.6%	.0%	1.6%	9.5%	3.2
TV/Radio	18.1%	4.4%	9.8%	3.8%	9.4%	9.5%	8.4
All	1.4%	10.6%	8.2%	9.4%	10.9%	11.0%	8.1

Source: Adapted from Rural Electrification Board.

The second, from the Shahjalal University of Science & Technology in Bangladesh, documented that almost all of the households they surveyed found SHS units "financially attractable" and that in almost "all cases" units improved household incomes and living standards in rural communities.[86]

The third, from Australian researchers at Murdoch University and the University of Western Australia, noted that SHS had immediate financial benefits by displacing expensive monthly kerosene expenditures typically ranging from $5 to $14 (but even higher in some cases). It documented that one-third of households surveyed used their systems for lighting and the operation of televisions, but also that SHS use raised incomes and enhanced social status. As the authors stated:

> Almost all respondents agreed that using SHS had resulted in an increase in their quality of life. All respondents agreed that SHS increased time spent in relaxation and their ability to get together at night and enjoy the high quality (white and clear) light. The majority of the respondents (92 percent) reported that their systems had enabled them to use lights whenever they wanted to and to watch TV programs without interruptions by power failures and that they did not have to worry about charging their batteries in order to watch TV. They agreed that by watching TV or listening to the radio they had greater access to information and were more informed about weather (tornados, cyclones) and other natural disasters. Ninety-five percent of respondents agreed that their access to information through mobile phone or TV or radio had been improved by their SHS. They reported that they were able to charge their phones at home whenever they needed to use them. Most respondents (82 percent) also agreed that their SHS has increased their social status, stating that neighbors and relatives from other villages visited their houses more often to enjoy the clean lighting. They also claimed that their SHSs had increased the amount of time that they engaged in social activities. The capacity to increase income generation using SHS was also seen to be a major benefit of the SHS program... In this way, the SHS program has helped to reduce poverty by creating opportunities for new income earning activities, such

as mobile phone charging shops, providing neon light traps for attracting and destroying insects, and operating social TV halls.[87]

Consequently, SHS adoption is not driven merely by economic incentives to improve income, but by "non-income" factors that can play a "significant role" in household adoption, factors such as families desiring a clean home free of kerosene fumes, wishing to charge mobile phones, or hoping to impress their neighbors. wanting to give children better-quality light to study at night.[88]

Better technology

A final benefit relates to better technology and domestic manufacturing. Near the beginning of the IDCOL program in 2003, all technology had to be imported from Japan, India, and China; there were few domestic manufacturers of components and none for panels or assembly. The market was also dominated by one company, Kyocera, which had an 80 percent share. Now, however, respondents reported that 30 suppliers of SHS components from 20 countries have facilities in Bangladesh, in addition to seven local manufacturers of batteries, charge controllers, and inverters. Tubular plate batteries and flat type batteries are now entirely made within Bangladesh, more than 80 percent of charge controllers are locally assembled, and three-quarters of the fluorescent lamps attached to SHS are now made within the country.

Kyocera still dominates the market for solar panels and components, but holds less than a 50 percent share; it must now compete with Tata BP, Solar World, Suntech, and others. One company, Electro Solar, has started assembling 5 MWp of solar panels in 2011. At the time of our field research three other assembling facilities have been approved and will likely start producing 15 MWp more of panels by early 2013. One solar vendor said that "the IDCOL program has opened the door to a robust domestic manufacturing capability for SHS panels and components." The result has also been improved quality and lowered costs, with solar panels now costing about $1 per installed Wp.

Challenges

Notwithstanding the benefits of the IDCOL program, it also faces five ongoing challenges: (1) a trust gap, (2) affordability for the poorest customers, (3) aggregate capacity, (4) technical barriers, and (5) natural disasters.

Trust gap

The first relates to a continuing trust gap among end-users and politicians about SHS. One respondent from an NGO noted that "most Bangladeshi families" do not trust the technology, "with some clients actually visiting demonstration facilities two to three times before they are convinced a SHS will work as we claim ... the only way they are convinced is by showing them, often repeatedly."[89] Another, from civil society, talked about a cultural barrier whereby components and panels produced in Bangladesh and China are perceived as less valued than Western or Japanese systems. As they noted:

> Chinese electronics, and even Bengali technology, are widely believed to be of universally poor quality. It is a challenge trying to get people here to buy a Chinese made system or a domestically manufactured system, people are unwilling to invest so they end up preferring the more expensive American, German, or Japanese systems, that's how negative the reputation of local technology is.[90]

To our knowledge these perceptions are unfounded, but it is possible that in some cases differences in quality do exist. Respondents added that such gaps of trust exist not only for users but also for political leaders.

Affordability and rising prices

Even though IDCOL has tried to target use among families most in need, one respondent from the private sector critiqued that "SHS for all intents and purposes are still for upper and middle class rural houses, or joint families or big families that have enough people and resources to afford the down-payment, SHS are still beyond the reach of the poorest in Bangladeshi society." Others from various NGOs noted that "there are more poor people in Bangladesh than in all of Sub-Saharan Africa, these people cannot even afford a light bulb, how are they supposed to afford a SHS?," that "SHS is a costly technology, well above what poor Bangladeshi family can make in a year," and that "SHS are a very expensive item, only the richest people in a poor community can afford it."[91] Still others argued that SHS sit near the top of the income bracket, with solar lanterns and improved cookstoves in use for poor households but SHS and biogas units in use "only for wealthier families."[92] This could be why one of the SHS homes we toured near Singair was occupied by a large family of 30 that had many livestock – making them comparatively wealthy. As one critic of the IDCOL program commented, "by my count 40 percent of Bangladeshis in rural areas still cannot afford SHS being subsidized through IDCOL, they are simply too poor, and IDCOL seems more interested in numbers and making returns on investment than in helping poor people."[93]

Aggregate installed capacity

While the number of systems installed is impressive, one respondent from a government ministry noted that "from a larger picture SHS have yet to make a true dent in the national supply of electricity, we're talking less than half of one percent of total demand for power."[94] And, although such criticism may be unfair, since the IDCOL program never intended to replace the grid, another respondent from the private sector stated that "the hundreds of thousands of SHS being installed in Bangladesh sounds impressive, but... I don't see SHS replacing centralized generation anytime soon."[95]

Moreover, some experts in Bangladesh have accused the focus of SS on off-grid solar as "picking the wrong technology," and suggested that instead the money should have gone towards commercial-scale, grid-connected renewables. One recent study conducted jointly by the Center for Development Research in Germany and the Islamic University of Technology in Bangladesh partially confirms this point.[96] It noted that the potential for grid-connected solar PV systems was so large throughout 14 "widespread" locations that Bangladesh could install a whopping 50,174 MW of it – to put this figure in perspective, Bangladesh's current grid operates about 5,400 MW. Not only do such centralized solar PV facilities have the potential to expand existing electricity capacity by roughly a multitude of ten, but they can also do a number of things off-grid systems cannot do. The larger-scale grid-connected solar facilities, the study notes, can reduce energy and capacity losses throughout the entire distribution network, and avoid the need for costly transmission and distribution upgrades.[97]

Technical barriers

Larger-capacity batteries with longer lifetimes, improved warranties on imported solar panels, efficient recycling and disposal of panels, and their operation during longer periods of reduced sun during the monsoon season were all mentioned as current areas of concern during the field research undertaken by the author in Bangladesh, along with a dearth of opportunities for local research institutes to become more involved in SHS research and development. Many of these barriers were recently confirmed by an independent assessment of the solar market in Bangladesh by the Department of Electrical and Electronic Engineering at the Ahsanullah University of Science and Technology.[98] That study, conducted in 2012, documented lack of national quality standards for solar equipment, the limited nature of local manufacturing and assembly of SHS units, and the limited capacity to design, operate, and install solar systems as significant "technical barriers."

Natural disasters

Floods, landslides, and tsunamis are among only a few of the recent events that have directly destroyed SHS, as well as cultivated and arable land, thereby reducing the capital villagers have available to make the down-payment for systems. One respondent from the government noted that "Cyclone Sidr in 2007 destroyed hundreds of SHS installed on the coast in a matter of hours and flooded thousands of hectares of arable land. People there still haven't fully recovered and investments in SHS in the coastal areas has since plummeted."[99] Another, from a multinational institution, remarked that "the sheer uncertainty and frequency with which natural disasters strike Bangladesh complicates any decision to invest in a SHS. People are too worried it will just literally blow away" (though better mounting and the removal of panels before severe storms potentially mitigate these problems, and IDCOL has created a SHS Disaster Fund so that victims of disasters do not have to pay for the rehabilitation of systems entirely on their own).

Conclusion: lessons and implications

Perhaps the simplest lesson from IDCOL's program is that investing in renewable energy systems, or expanding access to modern energy, can spill over into enhanced development and the achievement of higher living standards, lowered fuel consumption or fuel prices, improved technology, better public health, and reduced greenhouse gas emissions. Despite the mitigating effects of a persistent trust gap, affordability concerns, inability to displace large power plants, and natural disasters, the IDCOL's Solar Home Systems program has disseminated 1.42 million units servicing more than five million individuals. It has created thousands of local jobs, improved the distribution and reliability of energy services such as lighting and entertainment, displaced the use of more expensive kerosene, generated income for thousands of communities, and supported the establishment of a domestic manufacturing center for solar panels and SHS components.

Second, the SS program reiterates the importance of finding "appropriate technology" for communities and developing countries. In his widely acclaimed *Small is Beautiful: Economics as if People Mattered*, economist E. F. Schumacher wrote in 1973 that large, expensive technological projects continue to fail (or bring about unintended or undesirable consequences) because they are being implemented on the wrong scale. Schumacher developed the term "appropriate technology" to encompass his belief that technologies must: (a) support local economic growth within a given community; (b) create independence from outside sources of knowledge and capital; (c) employ the simplest production methods available; and (d) use local materials that minimize harm to the social and natural environment.[100] The SHS being promoted by SS meet these criteria, for they are matched in quality and scale to household needs for lighting and income generation, they are simple to operate and repair, and they minimize dependence on outside expertise and intellectual property.

Third, the SS experience suggests a set of design principles for other energy access programs. By the active promotion of high-quality technology and specifications through rigorous inspections and technical standards, consumers receive systems they can generally trust. Choosing high-quality participating organizations, those that show they can make a return on investment, collect payments, and build a national community-based network to assist with implementation and commercialization, ensures that the program operates smoothly. Specifically targeting services for both extremely poor households and for income-generating activities like pumping water, selling items, and sewing at night as well as enabling the use of mobile phones, televisions, and radios ensures that they fight poverty in addition to improving quality of life. The innovative financial model of the program, which relies on a phased reduction of grants combined with subsidies, low-interest loans, and a down-payment by households themselves, ensures that costs are shared, and that multiple stakeholders become vested in the program. Taken together, IDCOL's experience implies that, if one follows all of these lessons, the result is more educated, affluent, resilient communities.

If IDCOL can achieve such rapid diffusion of SHS in Bangladesh – a country that since 1971 has seen a war, multiple famines, disease epidemics, killer cyclones, massive floods, military coups, political assassinations, and extremely high rates of poverty and deprivation – it makes one wonder: where can't it be done?

8
Intragenerational Equity and Climate Change Adaptation

Introduction

Our global energy system is unjust because it emits greenhouse gases into the atmosphere, which then damage communities and countries that, given their primarily agrarian economies, contributed the least to those emissions.[1] Even worse, these emissions will likely harm existing and future generations with a collection of grave consequences such as rising food insecurity, the proliferation of climate refugees, and an increased frequency and severity of natural and humanitarian disasters. The buffering capacity of the planet requires that we radically reduce our emissions at the same time that the energy needs of the world are crying out for a vast increase in the distribution of energy (see Chapter Seven for more on that). Fortuitously, community-based adaptation measures provide developed and developing countries alike a way out of this quagmire. The Global Environment Facility's (GEF) Least Developed Countries Fund (LDCF) offers an exemplary model for how these adaptation projects can be implemented. It has, as of late 2012, disbursed $602 million in voluntary contributions in support of 88 adaptation projects across 46 countries.

Intragenerational equity as an energy justice concern

Because climate change is such a wide-ranging and complex threat, it cuts across multiple justice concepts and dimensions. The impacts of climate change will be distributed unevenly due to both physical processes and the different adaptive capacities of communities and countries; also, historically only a small group of countries is responsible for the largest chunk of these emissions.[2] Justice expert Gordon Walker writes that:

> Climate change makes the most persuasive case for a justice framing. With climate change we are confronted with evidence of patterns of inequality and claims of environmental injustice that span the globe,

that permeate daily life and which pose threats to the current and future health and well-being of some of the poorest and most vulnerable people in the world. Climate change demands more than ever that we think rationally about how things interconnect, about who benefits at the expense of others, and about the spatially and temporally distant impacts of patterns of consumption and production. The consequence is that, for many already economically, politically, and environmentally marginalized people, climate change presents compounding forms of injustice.[3]

This makes the issue of climate change about fairness. As University of East Anglia climate expert W. Neil Adger and his colleagues write, "fairness is essential to reaching any meaningful solution to the problem of climate change during this century."[4] But fairness based on what, and to whom?

Modern justice theorists advance at least two interrelated answers: an argument about future generations, and an argument about human and subsistence rights. First, climate change raises justice concerns for future generations in a variety of ways. Failing to mitigate emissions today will inflict actual harm on future people when those emissions produce dangerous changes in climate.[5] The climate-related impacts of past and current greenhouse gas emissions could last longer than Stonehenge, time capsules, and perhaps even high-level nuclear waste. For each ton of carbon dioxide we leave in the atmosphere today, one-quarter of it could still be affecting the atmosphere one thousand years from now.[6] Once emitted, a ton of carbon dioxide takes a very long time to process through the atmosphere – according to the latest estimates, one-fourth of all fossil-fuel-derived carbon dioxide emissions will remain in the atmosphere for several centuries, and complete removal could take as long as 30,000 to 35,000 years.[7] Put another way, the climate system is like a bathtub with a very large tap and a very small drain.[8]

Consequently, "future generations will be more severely damaged by climate change than present generations – indeed, they will be its greatest victims, especially in the relatively near future before physical and psychological adaptations can set in for the lucky."[9] University of Tennessee philosopher John Nolt has gone so far as to frame the current situation as equivalent to the "enslavement" of future generations. He writes that "our emissions of greenhouse gases constitute unjust domination, analogous in many morally significant respects to certain historic instances of domination that are now almost universally condemned, and, further, no benefits that we may bequeath to the future can nullify the injustice."[10]

Second, like the situation of energy access and energy poverty discussed in Chapter Seven, climate change raises justice issues on human rights grounds. Justice theorist Henry Shue has argued compellingly that, if physical security is a basic right, then so are the conditions that create it, such as employment, food, shelter, and also unpolluted air, water, and other environmental goods, something he calls "subsistence rights."[11] The implication is that such people

Table 8.1   Criteria and indicators for Shue's "Standard of Decent Living"

Basic Goods	Energy Service(s)	Standard for Decent Living
Food	Cooking energy, methane	Adequate nutrition, 2 MJ/cap/day
Water/Sanitation	Heat for boiled water	50 liters potable water/month
Shelter	Floor space, lighting, space conditioning	10 square meters of space, 100 lumens per square meter light, 20 to 27 degrees C temp
Health care	Electricity	70-year life expectancy
Education	Lighting and electricity	
Clothing	Mechanical energy for weaving	
Television	Electricity	~100 kWh per month for all household appliances
Refrigerator	Electricity	
Mobile phone	Electricity	
Mobility	Personal vehicle	Motorized transport

Source: Adapted from Shue.

are therefore entitled to a certain set of "goods" that enables them to enjoy a basic minimum of well-being, shown in Table 8.1; included in this set of goods is the right to "subsistence emissions." As Shue puts it, "Basic rights are the morality of the depths. They specify the line beneath which no one is allowed to sink."[12]

In sum, these complementary notions of protecting future generations and ensuring subsistence rights mean that "distance makes no moral difference in our globalized world; individual high emitters have a duty to reduce their emissions, wherever they are."[13] It means, moreover, that when some people have less than enough for a decent human life, and other people have more than enough, an adequate minimum must be set for those who need to meet a basic standard of living.[14]

Unfortunately, the pending impacts of climate change directly threaten our ability both to protect future generations and to meet our subsistence obligations. Though not a complete list, six climate-related impacts will likely be most severe: ocean acidification, more frequent and intense disasters, mass climate refugees, food production, disease epidemics, and shortages of water.

Ocean acidification

Due almost completely to emissions of carbon dioxide, the acidity of the oceans has increased about 30 percent since the time of the Industrial Revolution, the greatest rate of increase in the past 55 million years – posing serious threats to countries in Asia and Africa that mostly depend on fish for their food. At the base of the marine food chain, acidification is rapidly depleting algae and plankton. Increased acid is bleaching and damaging coral reefs. For example, clown fish are especially susceptible to acidification as they will lose their ability to "smell." Acidification is interfering with the reproductive processes of brittle stars, which is in turn shrinking stocks of herring. It is also causing a decline in the levels of aragonite and calcium carbonate, key to almost all marine skeletons and shells.

Rather than being limited in scope, the threats from acidification are global, with the greatest degree of acidification in the Atlantic, north Pacific, and Arctic seas, each a crucial summer feeding ground for billions of organisms.[15] Recent scientific studies warn that climate change will likely lead to numerous local extinctions and drastic species turnovers (invitation to and extinction from an area) affecting more than *60 percent* of all marine biodiversity, as well as declines in the vitality of coral reefs due to bleaching, diseases, and tropical storms – with roughly *one-third* of all coral reefs at risk of becoming extinct.[16] If allowed to run its course, such acidification could turn the shining sea into a "carbon cesspool."[17]

Natural and humanitarian disasters

Climate change is increasing the frequency and severity of natural and humanitarian disasters. Global economic damages from natural catastrophes, most of them related to climate change, have doubled every ten years and reached about $1 trillion over the course of the past two decades. Hurricane Sandy, which recently flooded parts of New Jersey and New York in October 2012, caused up to $50 billion in damages to New York (and that's excluding the destruction in its wake across the Bahamas, Cuba, Dominican Republic, Haiti, Puerto Rico, and in other parts of the U.S.).[18] Annual weather-related disasters have increased by a factor of four from 40 years ago, and insurance payouts have increased by a factor of 11, rising by $10 billion each year for most of the 1990s.[19] In Colombia, for example, changing precipitation patterns, more extreme weather events, hurricanes and flooding, stronger cycles of El Nino and La Nina, and increases in sea level both threaten coasts and will challenge the vitality of low-lying urban centers.[20] In Bolivia and parts of Latin America, flooding from storms is expected to contribute to landslides that result in thousands of deaths and the spread of diseases.[21] In the Maldives, about half (44 percent) of all human settlements and 70 percent of all critical infrastructure are within 100 meters of the sea. These settlements are already at risk from rising sea levels, storms, and floods. Severe weather events from 2000 to 2006 flooded 90 inhabited islands at least once and 37 islands repeatedly. Sea swells in 2007 inundated 68 islands in 16 atolls, destroyed 500 homes, and necessitated the evacuation of 1,600 people.[22]

Mountainous areas around the world, from the Alps in Europe to the Himalayas in Asia, also face the aggravated risk of glacial lake outburst floods – when glaciers melt faster than expected and produce massive, spontaneous releases of water capable of killing thousands of people and destroying entire cities. The United Nations Environment Program and the International Center for Integrated Mountain Development have identified no fewer than 24 high-risk glacial lakes near Bhutan and Nepal.[23] Melting glaciers will flood river valleys in Kashmir and Nepal to the point where 182 million people could die of the resulting disease epidemics and starvation.[24] In Africa, rising sea levels could destroy as much as 30 percent of the continent's entire coastal infrastructure.[25]

The United States Department of Defense has simulated the probable effects of climate change and has begun preparing for a future world where droughts and periods of extreme heat increase in the Southwestern United States and Mexico; where the intensity of hurricanes increases on the US coast and in the Caribbean basin; where ice storms become more difficult to deal with in New England and Eastern Canada; where large mudslides and flooding occur in Central America; where massive wildfires cause deforestation, flooding, and siltification not only in California, Washington, and Canada, but also in Argentina and Brazil; and where the Philippines, India, Bangladesh, Vietnam, and China respond to typhoons and cyclones that severely damage coastal cities.[26]

Food security

Climate change has grave implications for the production, processing, and distribution of food, with particularly severe impacts in Africa and Asia. According to one study published in the *Lancet*, by 2080 as many as 40 least developed countries with a total population of three billion people could lose 20 percent of their cereal production. Alterations to the ranges of agricultural pests and diseases with warming winters could cause infestations of locusts, whiteflies, and aphids that could create "extensive losses of crop yields." Over the past three decades precipitation across the Sahel in Africa has declined by 25 percent, contributing to hunger and malnutrition in the Niger Delta, Somalia, and Sudan. Some experts anticipate severe climate-induced shortages of food in Angola, Burkina Faso, Chad, Ethiopia, Mali, Mozambique, Senegal, Sierra Leone, and Zimbabwe that will starve 87 million people.[27] Another study warned that as many as 75 to 250 million people in Africa could be exposed to increased water stress by 2020 as yields from rain-fed farms fall by 50 percent.[28]

Countries in the Asia Pacific will also be hit hard. Some states, such as Maharashtra, India, are projected to suffer greater drought that will likely wipe out 30 percent of food production, inducing $7 billion in damages among 15 million small and marginal farmers.[29] In India as a whole, farmers and fishers will have to migrate from coastal areas as sea levels rise and they confront heat waves lowering crop output and manage declining water tables from saltwater intrusion.[30] In China, higher temperatures and increased evaporation rates for soil are expected to result in a ten percent overall increase in the water needed for agriculture, and farms will likely become more vulnerable to insects and pests, resulting in declining yields.[31]

In Laos, the government anticipates that almost half (46 percent) of the rural population will risk food insecurity due to loss of access to farmland and natural resources caused by a combination of flooding, droughts, and rising prices.[32] In Bhutan, farmers have already reported instabilities in crop yields, losses in production, declining crop quality, and decreased water available for farming and irrigation. Moreover, they have documented loss of soil fertility from erosion and runoff, delayed sowing of crops due to premature frost, and outbreaks of new pests and diseases.[33] In Bangladesh, home to 150 million people, higher temperatures and changing rainfall patterns, coupled with increased flooding and rising salinity in the coastal belt, are likely to reduce crop yields and crop production, taking their toll on food security. Some studies even calculate the likelihood of a 17 percent loss in overall rice production and as much as a 61 percent decline in wheat production in the next few decades; they caution that any positive increases in yield will be more than offset by moisture stress.[34]

Human health and diseases

The World Health Organization believes that climate change has already killed 150,000 people in 2000 and subjected a further 5.5 million people to years of lost life due to debilitating diseases; most of these instances have been in the developing world. More worryingly, the WHO projects at least a *doubling* in deaths and burden in terms of life years lost by 2030 due to heat-related illnesses, illnesses from floods, droughts, and fires, changing vector patterns, and loss of biodiversity.[35] In China, climate change will in all likelihood alter disease vectors and create conditions for pandemics, as increases in temperature and decreases in availability of water expand the range and frequency of malaria, dengue fever, and encephalitis.[36] In the Maldives, waterborne diseases such as shigella and diarrheal diseases have also become more pronounced in children under the age of five, spread by an increase in flooding. Indirectly, climate change has contributed to malnutrition and limited the accessibility and quality of health care, with storms and floods making it more difficult to distribute food or transport patients to doctors.[37] In low-lying river deltas throughout the world, flooding and cyclones will directly affect health and nutrition by causing physical damage and disruptions in the supply of food and basic services, and indirectly by spreading waterborne diseases and creating prolonged periods of malnutrition. During the monsoon season in 2004 in Bangladesh, for example, flooding placed 60 percent of the country under a solid pool of water mixed with industrial and household waste. More than 20 million people suffered shortages of water, skin infections, and communicable illnesses.[38]

Water quality and availability

The water-related impacts from climate change will probably be just as egregious, encompassing reduced access to freshwater, less water for irrigation, less drinking water, and improper sanitation. Changes in rainfall, snowfall, snowmelt, and glacial melt could put 40 percent of the world population at risk – since they depend on mountain glaciers for their water supply. Several of the world's major rivers, including the Indus, Ganges, Mekong, Yangtze, and Yellow, start from glaciers. Storm surges also threaten to contaminate water with saltwater.[39] By 2080 increased floods, droughts, and storm surges could all lower water availability and quality and affect 1.5 billion people.[40]

Climate refugees

Global climate change's impending threats will push many families out of their homes. These climate refugees must relocate due to the many impacts discussed above. According to the Environmental Justice Foundation, "[e]very year climate change is attributable for the deaths of over 300,000 people, seriously affects a further 325 million people, and causes economic losses of $125 billion."[41] A separate study calculated that by 2050 more than 200 million people could lose their homes due to climate change.[42]

While environmental calamities have been common throughout the ages, the world's population intensifies the scope of the climate refugee problem; as the *New York Times* explains, "with the prospect of worsening climate conditions over the next few decades, experts on migration say tens of millions more people in the developing world could be on the move because of disasters."[43] Similarly, small island developing states such as the Maldives and the Seychelles could be completely submerged within 60 years if sea levels continue to rise. The Republic of Kiribati, a small island country in the Pacific, has already had to relocate 94,000 people living in shoreline communities and coral atolls to higher ground.[44] The Republic of Maldives could lose 80 percent of its land due to rises in sea level and has already started purchasing land in Sri Lanka for its climate refugees.[45]

Community-based adaptation

Community-based adaptation measures offer perhaps the best tool enabling us to meet our climate change justice obligations. The term "adaptation" describes adjustments in natural or human systems in response to the impacts of climate change.[46] The Intergovernmental Panel on Climate Change defines "adaptive capacity" as the ability of a system to adjust to climate change (including climate variability and extremes) to moderate potential damages, to take advantage of opportunities, or to cope with the consequences.[47] Adaptation is "community-based" when implemented by local stakeholders, and it is sometimes called "anticipatory adaptation" when it tries to preempt particular risks.[48] Most adaptation efforts are targeted towards enhancing "resilience," the amount of disturbance a local system, climatic or social, can absorb and remain within the same state.[49] Table 8.2 illustrates three of the most salient types of resilience.

Table 8.2 Dimensions of resilience and adaptive capacity

Type of Resilience	Explanation	Dimensions
Infrastructural	Refers to the assets, infrastructure, technologies, or "hardware" in place to ensure the delivery of services that could be disrupted by climate change (such as electricity or water)	Resilient infrastructures tend to encompass relevance, flexibility, and diversification
Institutional	Refers to the endurance of an institution or set of institutions, usually government ministries or departments, in charge of planning and community and infrastructural assets	Resilient institutions are strong; they can cope with new stresses and changes and maintain their core function and purpose. They tend to have permanence, rapidity, and legitimacy
Community	Refers to the cohesion of communities and the livelihoods of the people that compose them	Resilient communities tend to possess ownership, wealth, education, and access to knowledge and education that enable them to make decisions and respond to climate-related challenges

As the case study of the LDCF below shows, adaptation efforts are necessary if communities are to respond to drastic changes in climate once tipping points, such as acidification of the ocean, alteration of the Gulf Stream, or thawing permafrost, are crossed, and adaptation can also have a high relevance regarding slow or gradual changes in climate.[50] Furthermore, adaptation efforts tend to be "win-win situations," for they not only improve resilience to climate change but often spill over into ancillary benefits such as economic stability, improved environmental quality, community investment, and local employment.[51]

Case study: The GEF's Least Developed Countries Fund

Established in 2001, the GEF's LDCF was created exclusively to help the poorest countries in the world prepare and implement National Adaptation Programs of Action (NAPAs) to reduce the pending impacts of climate change. Currently one of the world's largest funds for climate adaptation, the GEF has so far leveraged $602 million in voluntary contributions to support 88 adaptation projects in 46 countries (as of November 2012), projects implemented in tandem with partner agencies including the World Bank, United Nations

Development Program, and Food and Agriculture Organization. As the GEF explains, the LDCF is "seminal in climate change adaptation finance" and is the "first and most comprehensive adaptation-focused program in operation for least developed countries."[52]

The Fund is special because, as the name implies, it is dedicated almost entirely to the 48 countries belonging to the group of "least developed," meaning they have low incomes (less than about $900 per capita per year), weak human assets, and high economic and social vulnerability. Least developed countries lack the requisite capacity to implement adaptation projects. While the city of Perth in Western Australia can build a desalination plant to offset losses in water due to declining precipitation and increasing drought; planners in the Netherlands can construct dikes, dams, and floating houses to cope with increased flooding and rises in sea level; and the city of London can invest in a Thames River barrier system to better respond to floods, some of the world's poorest areas have no resources to implement adaptation projects on their own.[53] Least developed countries depend on climate-sensitive sectors such as agriculture, tourism, and forestry, meaning changes in temperature and precipitation and extreme weather events affect them more viscerally than others. They are also, for a variety of geographic and economic reasons, located in regions at the greatest risk of rising sea levels, deteriorating ecosystem services, social tensions, and the creation of environmental refugees.[54]

History

The LDCF arose out of the Seventh Session of the Conference of the Parties to the United Nations Framework Convention on Climate Change (COP7), held in Marrakesh, Morocco, in 2001. The GEF was placed in charge of this "financial mechanism" for climate change and the World Bank was established as its "Trustee." As Table 8.3 shows, the LDCF is one of five major multilateral funds for adaptation projects.[55] The LDCF has a governing body which meets twice a year. The LDCF supports two key activities: the preparation of NAPAs, policy documents assessing climate-related risks for particular countries; and the implementation of adaptation projects prioritized according to each country's specific NAPA. All least developed countries are eligible for the fund – it operates according to the principle of "equitable access" rather than "first come, first served," though proposals are formally evaluated based on their country of origin, conformity with existing national policies, and institutional support, among other criteria.[56]

Since its creation, as of June 2012 – the last time formal numbers were released – the LDCF has funded the completion of 48 NAPAs and the implementation of 74 projects and one program across 44 countries, totaling $334.6 million and leveraging $1.6 billion in co-financing (rising to $602 million, 88 projects, and 46 countries by the end of the year).[57] As Figure 8.1 illustrates, these projects have focused on reducing vulnerability across a

Table 8.3 Funds for adaptation under the United Nations Framework Convention on Climate Change Regime

Fund	Created under	Global environmental benefits	Beneficiaries	Funding sources
GEF Trust fund	UNFCCC	Incremental cost to achieve global environmental benefits	Developing countries	GEF
GEF Strategic Priority for Adaptation (SPA)	UNFCCC	Incremental cost to achieve global environmental benefits	Developing countries	GEF
Special Climate Change Fund (SCCF)	UNFCCC	Additional costs of adaptation measures. Uses a sliding scale	Developing countries	Developed countries discretionary pledges
Least Developed Countries Fund (LCDF)	UNFCCC	Additional costs of adaptation measures. Uses a sliding scale	Least developed countries	Developed countries discretionary pledges
Adaptation Fund (AF)	Kyoto Protocol	No	Developing countries	Share of proceeds from CDM; other sources

Source: Adapted from United Nations Framework Convention on Climate Change and Grasso. CDM refers to the Clean Development Mechanism.

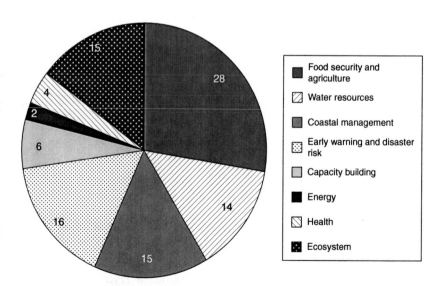

Figure 8.1 Development sectors prioritized in National Adaptation Programs of Action (as of June 2012)

Source: Global Environment Facility.

variety of sectors, including early warning and natural disasters, agriculture, and fragile ecosystems. Figure 8.2 documents how, as of June 2012, some $537 million had been pledged to the LDCF from countries such as Germany, the United Kingdom, the United States, and Sweden, among others.

Because covering all 88 projects would make this chapter excessively long, it focuses mostly on four major efforts being implemented in Asia and summarized in Table 8.4: coastal afforestation in Bangladesh, glacial flood control in Bhutan, agricultural production in Cambodia, and coastal protection in the Maldives.

Bangladesh is prone to a multitude of floods, droughts, tropical cyclones, and storm surges. Fifteen percent of its 162 million people live within 1 meter's elevation from high tide. In 1991, a particularly devastating cyclone with winds stronger than 200 kilometers an hour and a tidal surge of 6 meters claimed 140,000 lives and induced $240 million in damages.[58] Such climatic vulnerabilities are only compounded by a high incidence of poverty and heavy reliance on agriculture and rural forestry. Rising sea levels place more than 40 million people at direct risk of saltwater intrusion of water supplies for drinking and irrigation, and the ever-present occurrence of floods from drainage congestion and severe storms.[59]

To respond to these threats, the Ministry of Environment and Forests is aiming to reduce the vulnerability of coastal communities to the impacts of climate change by carrying out LDCF-sponsored afforestation in four *upazilas* (translated as sub-districts) in the coastal districts of Barguna and Patuakhali (Western region), Chittagong (Eastern Region), Bhola (Central Region), and Noakhali (Central Region). Project managers selected sites on

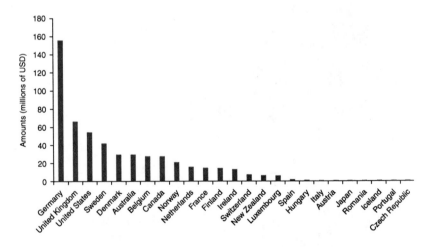

Figure 8.2   Least Developed Country Fund pledges by country (as of June 2012)
Source: Global Environment Facility.

Table 8.4 Examples of Least Developed Countries Fund projects in Asia

Sector/Type of Adaptation	Country	Budget	Duration	Primary Actors	Description
Coastal Afforestation	Bangladesh	$10.8 million	March 2009 to February 2013	United Nations Development Program; Forest Department at the Ministry of Environment and Forestry	Utilizes community-based afforestation, mangrove regeneration and plantation management, erosion prevention, and the deployment of coastal sediment barriers to reduce climate vulnerabilities in four upazilas in the coastal districts of Barguna, Patuakhali, Bhola, Noakhali, and Chittagong
Glacial Flood Control and Early Warning Systems	Bhutan	$8.3 million	March 2008 to February 2013	United Nations Development Program; Department of Geology and Mines at the Ministry of Economic Affairs; Disaster Management Division of the Ministry of Home and Cultural Affairs	Implements a disaster management plan and will demonstrate technologies available to reduce glacial lake outburst floods from the Thorthormi glacial lake in the Punakha-Wandgi Valley and Chamkhar Valleys
Coastal Protection	Maldives	$9 million	March 2010 to February 2013	United Nations Development Program; Ministry of Environment, Energy, and Water	Establishes a climate information system to collect and disseminate knowledge about climate vulnerability and designs a migration plan of Safer Islands to which communities at threat can relocate
Agriculture	Cambodia	$4.4 million	July 2009 to June 2013	United Nations Development Program; Ministry of Agriculture, Fisheries and Forestry; Ministry of Water Resources and Meteorology	Trains engineers in climate-resilient irrigation design (including reservoirs, irrigation canals, ponds, and dykes) and establishes a community-based climate information system on floods and droughts

Source: Global Environment Facility.

the basis of their projected vulnerability and also through public participation. The project has four primary components. The first is implementing interventions that generate income and couple afforestation with community livelihood. The second is enhancing national, sub-national, and local capacities of government authorities and sectoral planners so that they better comprehend climate risk dynamics in coastal areas and implement appropriate risk reduction measures. The third is reviewing and revising coastal management practices and policies. The fourth is developing a functional system for the collection, distribution, and internalization of climate-change-related data.

In Bhutan, the acceleration of glacial melting has compounded the risk of Glacial Lake Outburst Floods (GLOFs). Glacial lakes there hold tens of millions of cubic meters of water and can release high volumes in minutes, devastating valleys and communities downstream. Major sectors of the economy involve agriculture, livestock, and forestry, but these have become situated in close proximity to flood paths. One inventory identified 25 glacial lakes at "high risk" of a GLOF, with 12 located in the Pho Chhu and Chamkhar Chu sub-basins, home to more than 40 Bhutanese villages and towns with tens of thousands of residents.[60]

In response, the government launched the LDCF-funded GLOF project to tackle disaster risks. It has three primary components. A component focused on lowering of lake water levels is being undertaken by the Department of Geology and Mines (DGM) to reduce the risk of GLOFs at two glacial sites in the Himalayas. So far mitigation work by DGM has focused only on one lake, Thorthormi, where it is aiming to reduce the lower lake's water level by 5 meters, enough to eliminate hydrostatic pressure on its unstable moraine dam. An early warning component is being led by the Department of Energy, and a third community awareness component is attempting to increase knowledge about climate change among community leaders and rural policymakers.

In Cambodia, droughts and floods have already caused substantial human and crop losses and are widely viewed as a prelude to more extreme weather. Rice, Cambodia's largest crop by volume and value, is forecasted to suffer yield losses of 5 percent over current levels in 2020 under IPCC scenarios. Annual rainfall is projected to increase in some areas, but, when coupled with increased variability and ambient temperatures, yield losses will worsen through 2080 and potentially turn Cambodia into a net rice importer.[61]

Planners there are, therefore, focusing on building adaptive capacity with LDCF resources for water management and agriculture. They are enhancing the ability of local government and communities to integrate long-term climate risks into policy and decision-making related to subsistence farming and rice paddy production. Many of the communities living in the targeted districts in Preah Vihear and Kratie practice subsistence farming and are reliant on agriculture for their livelihoods. Adaptation efforts are focusing on educating these farmers and local leaders about climate change, and also strengthening infrastructure such as irrigation channels and ponds.

Geographic and geophysical traits, such as small size, low elevation, narrow width, and dispersed nature of coral islands and reefs, make the Maldives especially vulnerable to rainfall flooding and ocean-induced flooding. About half the country's human settlements are within 100 meters of the shoreline, along with almost three-quarters of its critical infrastructure, including airports, power plants, landfills, and hospitals. The Maldives is the "flattest country on earth" and "extremely vulnerable" to climate change, so much so that 85 percent of its geographic area could be underwater by the year 2100 if sea levels rise under more extreme projections.[62]

The Maldivian government is thus using its LDCF money to integrate climate change risk management into formal planning processes. It is funding demonstration projects on four islands that promote a suite of different infrastructural improvements, including beach nourishment, coral reef propagation, land reclamation, and community relocation. The project is also creating "composite risk reduction plans" to be integrated with coastal protection and adaptation measures. Disaster risk profiles are being created for the four demonstration islands, revised and updated as scientific knowledge about climate change and sea level rise accumulates. These are to be synthesized into a national-level "multi-hazard early warning system."

Benefits

As this section documents, these four particular LDCF projects are producing four sets of distinct benefits: (1) strengthening nationally significant infrastructure, (2) enhancing institutional capacity and awareness, (3) improving community assets, and (4) producing benefits that exceed costs.

Strengthening infrastructure

Each of the four LDCF projects enhances physical and infrastructural resilience in some way. Bangladesh's coastal forest today is almost a monoculture of mangroves. These monoculture forests have a limited ability to mitigate the impacts of climate change as they have been prone to pest outbreaks, deforestation, and logging. Historically, Bangladesh had a 500-meter buffer of mangroves to reduce the shocks of incoming storms and monsoons, but this has now been reduced to 12 to 50 meters in most locations. Attacks by the stem borer pest have felled thousands of hectares, and illegal deforestation and logging have made matters worse. The project in Bangladesh addresses this problem and sponsors 6,000 hectares of community-based mangrove plantations, 500 hectares of non-mangrove mount plantations, about 220 hectares of dykes, and more than 1,000 kilometers of embankments. The Bangladesh project is also developing early warning information and disaster preparedness systems in vulnerable areas to protect at least 20 villages and towns.

In Bhutan, planners are improving early warning systems and draining glacial lakes. Previously, the Bhutanese Department of Energy managed only a single station in Thanza, which housed two people with a wireless radio set, a single satellite phone that monitored glacial lake water levels, and (in all likelihood) copious amounts of hot coffee. The problem is that the two people did not always report for work, have fallen asleep, and could have been killed by the GLOF itself. Under the project, the government will replace the manual system with an automatic one composed of gauges monitoring glacial lake bathymetry (depth) as well as sensors along rivers connected to automated sirens. The project will also eventually expand the automated warning system to cover more glacial lakes.

In Cambodia, infrastructural resilience will be improved by the construction and rehabilitation of retention ponds, canals, dykes, and reservoirs that, due to years of neglect, are currently in disrepair. Instead of rehabilitating these irrigation systems using design parameters derived from historical hydrological patterns, the project aims to integrate climate forecasts into their upgrading so that the infrastructure can withstand future climatic events such as droughts or floods.

In the Maldives, planners have moved away from exclusively building capital-intensive sea walls and tetrapods to bolster infrastructural adaptation by replenishing natural sea ridges, planting mangroves and vegetation on shorelines, and raising the height of water storage tanks so they are no longer susceptible to sea swells and saltwater intrusion. The government has started propagating new coral reefs around Thulusdhoo and Kudhahuvadhoo and adopting beach nourishment activities to mitigate flooding. Planners are deploying what they call "soft" adaptation infrastructure. As one government official told the author:

> The key to the [project] is moving beyond hard infrastructure to soft protection, using ecosystems and trees as measures to improve resilience that are cheaper, environmentally more sound, and longer lasting than their capital- and technology-intensive counterparts. The sea wall around Malé, for example, cost $54 million to erect, or $12.4 million per kilometer. The Maldives has 2,002 kilometers of coastline, which would make protecting them all with a seawall a monumental $24.8 billion enterprise. With the country's current annual GDP, it would take more than three decades to raise the funds for such a task, let alone build the sea wall. We've also got only $9 million in total to work with for the [project]. What are we going to do, build half a kilometer of sea wall with [our LDCF] money?

This comment implies that one of the more innovative ways the project strengthens resilience is by deploying smaller-scale, less capital-intensive "soft" measures such as planting mangroves or improving coastal vegetation, instead of building more of the "hard" and expensive seawalls like the one shown in Figure 8.3.

Figure 8.3 The $54 million tetrapod seawall surrounding Malé, Maldives

Enhancing institutional capacity

Adaptation efforts in our four LDCF projects prioritize not only infrastructure but also improving institutions and propagating standards of good governance. In Bangladesh, the government provides free training sessions for local level administrators in disaster management and also facilitates input from civil society and community members in the formulation of state and national policies and regulations.

In Bhutan, training for government planners is intended to build institutional capacity. The project has sponsored the training of geologists and employment for civil engineering work, and funded the creation of community-based disaster management committees, whose job it is to highlight hazards and form district disaster management teams at village levels.

In Cambodia, their LDCF project is encouraging community development plans based on long-term climate forecasts and scenarios, budgeting for water resources investments that are appropriate for the anticipated risks. In addition to devolving ministerial functions to local levels where possible, the project has shifted responsibility for planning onto community groups. The project thus empowers commune councils, farmer water user communities, and planning and budget committees to play a more active role in adaptation projects.

In the Maldives, institutional capacity is being strengthened through the training of government officials in risk analysis, hazard mitigation, and land use planning. By 2014 the goal is to train at least 12 senior decision-makers and planners from national ministries in Malé, as well as all senior decision-makers in four provinces and atolls. Part of this component involves participating with local island leaders to share knowledge and learn about local efforts at deploying some of the "soft" adaptation measures described above.

Improving community assets

Each LDCF adaptation project enhances community and social resilience. In many parts of the coastal forests of Bangladesh, the average annual per capita income is less than $130, a fraction of the national average, rendering people completely dependent on wetlands and coastal forests to meet their subsistence needs.[63] To counter this incentive to damage forests for their survival, the LDCF project is disbursing revenues to vulnerable coastal communities so that they can diversify income sources and occupational training. One especially innovative dimension of this component is its focus on the "Triple F" model of "Forest, Fish, and Food." The coastal communities most vulnerable to rising sea levels – the places where mangroves need to be planted and forests replenished – are also those where farming and forestry are the primary sources of income. The "FFF" model attempts to maintain community livelihood and adapt to climate change at the same time by integrating aquaculture and food production within reforested and afforested plantations.

In Bhutan, a community awareness sub-component is being implemented in Punakha, Wangdi, and Bumthang. Officials are creating a zoning map to mark several safe evacuation areas and extremely unsafe zones, and setting up emergency operation centers at district administration offices to enable them to better handle crises. Communities are being trained in their response to calamities and emergency situations using mobile phones and radio broadcasts in addition to traditional sounding gongs and bells from monasteries. Figure 8.4, for example, shows posters depicting first aid and emergency response techniques in the case of a GLOF. These efforts will give communities a better understanding of the risks and hazards surrounding GLOF occurrences. This information also enables communities to better plan for where to locate infrastructure, homes, and farmland.

In Cambodia, in addition to devolving ministerial functions related to adaptation efforts to local levels where possible, millions of dollars of funds have been transferred to fund the agricultural adaptation projects selected by village planning committees.

In the Maldives, planners are attempting to increase awareness of climate change in the outer atolls. One Maldivian official told the author that the project will "help decentralized adaptation investment planning so that each island decides what to spend its own budget on, therefore creating incentive for islands to 'pick best value for the money' so that they have resources left to improve community welfare in other ways." The program will also send "training teams" to remote islands to "create awareness among the community so that they can take stock of existing vulnerabilities and soft adaptation measures."

Figure 8.4 A Bhutanese Department of Disaster Management poster on glacial lake outburst floods

Positive cost curve

Though certainly simplistic, there are three ways that one can argue the LDCF has a positive cost–benefit curve. The first is based on its likely, positive impact in the future. The Asian Development Bank estimates that every $1 invested in climate change adaptation in 2010 could yield as much as $40 in economic benefits by 2030.[64] Presuming this to be the case for LDCF case studies – an admittedly crude way of calculating things – the $537 million so far pledged by the LCDF for adaptation projects around the world will culminate in $21.5 billion in economic benefits.

Second, though, is the cost of the LDCF's mitigation of emissions. Though adaptation rather than mitigation is its primary goal, its portfolio of projects actually mitigates emissions as an ancillary accomplishment. One study estimated that the GEF avoids or prevents carbon dioxide emissions from entering the atmosphere for a cost less than $2 per ton.[65] Noted economist Richard Tol has meticulously tracked the difficulties in ascertaining the marginal damage cost of carbon dioxide, but has synthesized data from dozens of reputable sources and concludes that it is somewhere between $14 and $93 per ton.[66] Economists working for the US federal government, hardly a source biased in favor of prudent climate policy, have similarly harmonized results from three integrated assessment models examining five socio-economic scenarios with three fixed discount rates and concluded that the cost of a ton of carbon dioxide ranges from $4.70 to $64.90 in 2010 and $11.20 to $109.70 for 2035.[67] Economists Frank Ackerman and Elizabeth Stanton from the Stockholm Environment Institute report that the real social cost of carbon could be as much as $1,000 per ton by 2050.[68] These varying figures mean that the GEF's actions displace carbon dioxide 2.35 to 500 times more cheaply than its actual cost.

A third way the LDCF has a positive cost curve is its ability to leverage more money than it spends. In late 2011, when the Australian Government was conducting an independent assessment of the LDCF, it noted that the LDCF leveraged $919 million in co-financing, more than $4.20 for each dollar contributed by the fund.[69] The most recent publicly available data from June 2012 reports $334.6 million spent and $1.56 billion leveraged in co-financing, meaning each LDCF dollar raised $4.66 towards climate change adaptation.

Challenges

Though these benefits are significant, the LDCF also faces four challenges: (1) insufficient and uncertain funding, (2) a convoluted management structure, (3) the complexity of adaptation projects within the context of LDCs, and (4) an inability to eliminate some of the most meaningful climate-related risks.

Insufficient and uncertain funding

Because the LDCF is supposed to prioritize "equitable access" for all participating countries, individual projects have a "ceiling" on the amount they support. For instance, from 2001 to 2006 the cap on LDCF projects was $3.5 million, in 2008 it was raised to $6 million, in 2010 it was increased to $8 million, and today it is $20 million. Though the LDCF has a mandate to finance the full additional cost of adaptation, without a requirement for matching co-financing, in practice the ceiling inadvertently requires hosting governments to co-sponsor projects, or find other institutions such as the United Nations Development Program (UNDP) or Food and Agricultural Organization (FAO) to "match" contributions. Moreover, because the LDCF is voluntary, it is only replenished when donor countries decide to be generous, making it difficult to accurately predict the amount of resources available to countries over long timeframes.[70]

Furthermore, the LDCF is clearly insufficient to ensure the implementation of all needed adaptation projects. As noted earlier, so far the fund has leveraged slightly more than $600 million, yet the immediate adaptation needs of LDCs total at least $3 billion.[71] In 2007, even when the LDCF is combined with three other large multilateral funds, the amount spent on adaptation equaled about $283 million pledged and $32.8 million disbursed, rising to $711 million spent in 2012 – as Table 8.5 shows.[72] This creates a "huge gap" with an estimated $10 to $100 *billion* in annual funding needed to prepare all developing countries for climate change.[73] Similarly, an assessment from the Potsdam Institute for Climate Impact Research, European Environment Agency, and other institutions calculated that at least $70 to $100 billion of investment will be needed per year for every year from 2010 to 2050 if adaptation needs are to be met.[74] As one recent independent evaluation put it, "the output of these funds falls far short of the estimated needs."[75]

Table 8.5   Multilateral adaptation funds disbursed (in millions of US dollars)

Fund	Goal	2007			July 2012	
		Pledged	Received	Disbursed	Pledged	Dispersed
Strategic Priority for Adaptation	Pilot projects that address local adaptation needs and generate global environmental benefits	50	28	14.8	–	50
Least Developed Countries Fund	Implementation of most urgent adaptation projects in LDCs, based on NAPAs	163.3	52.1	12	537	334
Special Climate Change Fund	Activities aimed at adaptation as well as three other purposes: technology transfer, economic diversification, and support in key sectors	70	53.3	6	240.68	162.24
Adaptation Fund*	Concrete adaptation projects in developing countries that are particularly vulnerable to the adverse effects of climate change	–	–	–	350	165
Total		283.3	133.4	32.8	1,127.68	711.24

*The Adaptation Fund was intended to be a three-year pilot program and expended all of its money by 2010.*Source*: Adapted from Global Environment Facility and Flam and Skjaerseth.

A final uncertainty relates to whether the LDCF will continue to exist in the face of the creation of a Green Climate Fund. During the Sixteenth Cancun Conference of the Parties (COP 16) in 2010, industrialized countries pledged to mobilize $100 billion per year by 2020 to address the climate change needs of emerging economies. At the center of this pledge is the Green Climate Fund (GCF), which has already raised $30 billion in "fast-track" financing from 2010 to 2012. If it reaches the $100 billion amount, the GCF will be equivalent to the cost of the entire four-year Marshall Plan to rebuild Europe after World War II.[76] With the creation of this fund, the LDCF has an unclear future. As one formal review of the LDCF put it, "it is recognized that what happens next to the LDCF depends to a large extent on the outcome of the negotiations on adaptation financing between parties to the UNFCCC."[77]

Convoluted structure

As well as its funding, critics have attacked the LDCF for having a convoluted management structure that has resulted in unnecessary delays for projects. Part of the reason is that the LDCF is administratively and legally "outside of the GEF Trust Fund."[78] This fundamental difference, however, means that the LDCF had to create an entirely separate management structure. During the Fourteenth Conference of the Parties (COP14) at Poznan, Poland, in 2008, some least developed countries "expressed their frustration" at this structure, at the speed with which projects were allocated funding, and at the "long and complicated" nature of implementing NAPAs.[79] Figure 8.5, for example, shows how complex a typical LDCF project cycle for the UNDP, one of ten implementing agencies, can become.

Though managers at the GEF have made various attempts to expedite the process and improve the efficacy of the LDCF, and not all delays in project implementation can be attributed to structural factors at the GEF, three independent evaluations suggest that problems still remain. The first, conducted by the UNDP in 2009, noted "justifiable dissatisfaction" among participants "concerning the lengthy time periods and complex procedures required to move from the NAPAs to concrete projects. In some cases, these have led to time lapses of several years before projects get off the ground."[80] That review noted, for example, that projects took an average of 471 days to begin due to "bottlenecks" and the "many stakeholders and consultations involved." It found that even the preparatory phase required "a lot of work" that ended up being "demanding" for country offices and cautioned that GEF requirements and project criteria were "complicated" and "poorly understood." It lastly noted that the co-financing requirement of the LDCF meant some countries did not have the resources needed to get projects commenced.[81] The second review, conducted by the Danish Ministry of Foreign Affairs in 2010, concluded that "in order for the LDCF to play a complementary role to the emerging other climate change financing mechanisms greater responsiveness and flexibility of procedures will have to be introduced to ensure

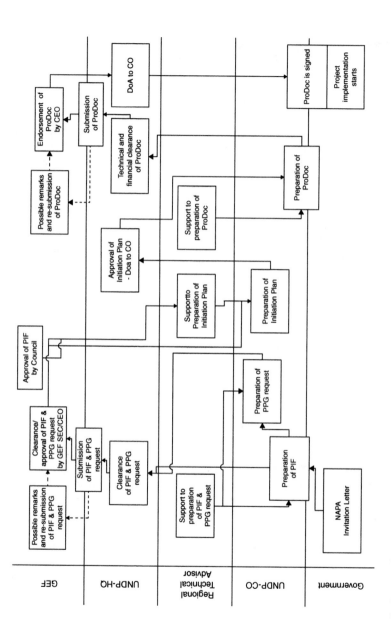

Figure 8.5 Process flowchart for a United Nations Development Program-Implemented Least Developed Countries Fund project
Source: United Nations Development Program.

lack of duplication and complementarity."[82] And the third review, from the nonprofit Climate Change Forum in 2010, criticized management structures at the LDCF that were "too complex," accused implementing agencies such as the World Bank as adding "further bureaucracy to the process," and concluded that "rules and structures make accessing funding difficult...and time-consuming."[83]

In the LDCF's defense, managers have attempted to address many of these concerns in earnest. One recent 2012 evaluation from the Australian Government noted that many of these problems have been addressed. It praised the LDCF for "successfully working with fragile states that are also least developed countries to develop the national adaptation programs of action" and noted that the majority of projects "have made satisfactory progress towards their development objectives." It commented that human resources were well managed, that monitoring for the program was "strong," and that "the Evaluation Office has made commendable efforts to improve and facilitate professional evaluation work in the GEF and to provide leadership, within the GEF partnership and internationally."[84]

Moreover, during the Eighteenth Session of the Conference of Parties in Doha, Qatar, in late 2012, the UNFCCC Secretariat commented that the average size of LDCF projects had grown from $3.5 million to $5.3 million.[85] It also noted that "there is evidence to suggest that LDCs have been able to learn from their initial experiences of NAPA implementation, and to scale up successful approaches and practices. Thanks to a streamlined project cycle, user-friendly guidelines for accessing resources, and enhanced communication between the GEF Secretariat and LDC stakeholders, proposals are being developed and processed faster."[86] As one example, the approval times for NAPA projects decreased from an average of 32 months to 12 months, with some taking as little as 75 days, and elapsed time between project approval and CEO endorsement for the most recent projects has shrunk from 17 months to 14 months.[87]

The complexity of adaptation

Though the LDCF has made serious progress in implementing scores of adaptation projects, their sheer complexity has nonetheless proven beyond the means of the technical and institutional capacity of many implementing stakeholders. In Bhutan, for instance, draining glacial lakes has proceeded much more slowly than expected. Thorthormi Lake is so remote that the nearest potential helicopter landing site turned out to be more than 90 minutes away by foot. The unstable terrain made the use of heavy machinery like excavators impossible, and site-to-site transport ended up damaging equipment and scientific instruments. Boulders and silt made it difficult to measure how quickly ice was retreating, and created a safety hazard as drifting icebergs and strong winds made bathymetric surveys dangerous; some boats actually capsized and dumped scientists into the freezing water.

Unpredictable weather played a part as well, with snow blocking the path to the site eight months of the year and storms, such as Cyclone Aila, preventing necessary equipment from reaching Bhutan as scheduled.[88] Heavy rainfall in 2009 also washed away several key bridges to the site, delaying work by days. Because of its finances, the Bhutanese government was unable to purchase high-resolution satellite imagery, nor does it own a single helicopter which would aid in monitoring. Under these conditions, project managers could only afford to pay a few hundred local volunteers (shown in Figure 8.6) who had to use shovels, spades, and a few jackhammers and chisels; no automated or heavy machinery was available.[89] As a result work has progressed "at a snail's pace, much slower than we had hoped."

In the Maldives, the "heterogeneity" or "specificity" of which adaptation measures work with each island has led to complications. As one adaptation practitioner told the author:

> The unique geography of Maldivian islands is a challenge when it comes to infrastructure, even softer adaptive measures. The needs of an elongated island on an outer atoll will differ greatly from those of a roundish island on an inner atoll. Patterns of sedimentation, the type and longevity of coral reefs, the socio-demographic composition of settled communities will all require different, site-specific options. There is likely not a "one size fits all" solution.

Broadening beyond the four demonstration islands is "essential" to truly protect the Maldives, yet "accomplishing this task in reality could prove difficult." In Bangladesh, despite the government's training efforts, capacity-building efforts have proceeded "weakly" and "slowly," and in

Figure 8.6 Bhutanese volunteers draining Thorthormi Lake in 2010

Cambodia the average government officer still possesses minimal knowledge about climate change and therefore may not see the necessity for adaptation efforts.

Inability to eliminate risks

The centerpiece of the LDCF is the creation of country-specific NAPAs, which represent a critical first step in implementing adaptation projects. These NAPAs, while useful tools, are ultimately only guideposts for how to prioritize adaptation investments; they do not directly provide the financing for those plans. As one GEF analyst put it, "the success of the NAPA process will largely be determined by how well it paves the way for scaled up investments in climate-resilient development in accordance with integrated, long term plans."[90] In other words, the presence of such plans is no guarantee their recommended measures will be implemented. It's hard to fault such countries for this shortcoming (of having a plan but not following it) since one study of adaptation planning in the United States, Australia, and the United Kingdom confirmed that these countries do the same.[91]

Even if countries were to fully follow the recommendations embodied in their NAPAs, however, there is no guarantee they can sufficiently increase resilience or lower climate-related risks. Consider the case with rising sea levels. Data from the IPCC suggest that by 2100 we could see temperature changes of 8.66 degrees Fahrenheit (4.81 Celsius) and that already, in 2012, temperatures have changed 1.03 degrees Celsius from preindustrial times (see Figure 8.7). Consequently, Figure 8.8 predicts an almost certain 1-meter rise in sea level by 2100. Under the most severe of these projections, if the Greenland Ice Sheet melts, sea levels could rise a stark 6 meters – enough to inundate almost all low-lying island states as well as coastal areas from San Francisco and New York to Amsterdam and Tokyo. Once a farfetched scenario, the destabilization of ice shelves and the sudden and unexpected collapse of the West Antarctic ice sheet now have scientists predicting "even greater likelihood of sea level rise in key regions."[92] Concentrations of carbon dioxide in the atmosphere could exceed 1,000 parts per million by volume by the year 2050 if trends continue.[93] Another 2011 synthesis of the scientific literature concluded that "there is now little to no chance of maintaining the rise in global mean surface temperature at below 2°C," and that "extremely dangerous climate change" will most certainly be incompatible with human and economic "prosperity."[94]

If sea levels rise as predicted under these scenarios, practically no amount of adaptation or investment in resilience can "save" countries such as Bangladesh or the Maldives. As one Bangladeshi government official told the author:

> The challenge Bangladesh now faces is to cope with changes in climate already happening every year. We are strengthening coastal embankments,

yes, but the intensity of erosion and frequency of storms are also increasing and I feel like we are often in a race against time where time is running out. We have developed saline-tolerant rice varieties but the concentration of salinity is going up. We can't keep on producing crops when land is flooded and water salty; it's practically not possible at the moment. Adaptation has its limits.

If the situation worsens, or if adaptation investments are not able to keep pace with vulnerabilities and risks, Bangladesh may have to switch to "retreat" measures such as forcibly relocating communities to higher ground.

Similarly, in the Maldives, such sea level rises would put the country "completely under water." Most islands are less than 1-meter high, meaning even small rises in sea level could subject the country to "regular tidal inundations."[97] These bleak and extreme projections may be why the Maldivian government is already relocating people to artificial islands, called

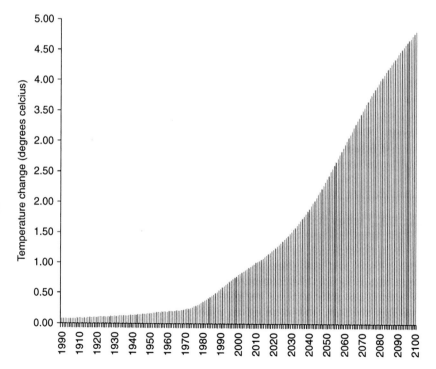

Figure 8.7 Expected global temperature change from preindustrial times (degrees C), 1900 to 2100[95]

Source: Adapted from International Energy Agency and Intergovernmental Panel on Climate Change.

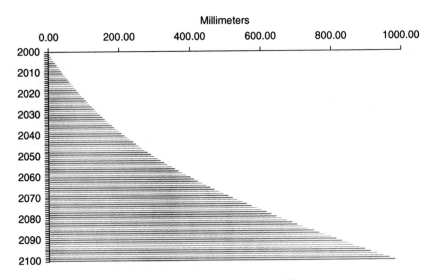

Figure 8.8   Expected sea level rise from 2000–2100 (mm)[96]

Source: Adapted from International Energy Agency and Intergovernmental Panel on Climate Change.

"designer islands." One such island, Hulhumalé in Malé Atoll, is set to house 100,000 people, many of them climate refugees, by 2030; construction is ongoing and it is currently home to 20,000 residents. The government also unveiled Dhuvaafaru Island in Raa Atoll in March 2009. Formerly an uninhabited forest, the entire island was raised and a new village built for the 4,000 survivors from Kandholhudhoo, an island destroyed by tsunami.

Conclusion: lessons and implications

The LDCF brings to light four salient conclusions.

First, it underscores the necessity of viewing resilience to climate change as multidimensional. As Table 8.6 summarizes, Bangladesh is not only sponsoring dykes and mangrove plantations, it is incentivizing agriculture and aquaculture to improve community income and training local officials. Bhutan is not only altering the physical shape of glacial lakes and rivers, building shelters, and creating an early warning system, but educating public and private leaders about emergency preparedness and climate risks. Cambodia is not only experimenting with crops and rehabilitating canals and ponds, but educating provincial officials and empowering local villagers to decide on infrastructure investments. Maldivian planners are not only thickening coastal vegetation and nourishing coral reefs, but decentralizing planning and disbursing funds directly to local communities so that they can decide what is best for them. Their efforts remind us that adaptation may

Table 8.6 Efforts in least developed Asian countries and their contributions to adaptation

Country	Infrastructural Adaptation	Organizational Adaptation	Social Adaptation
Bangladesh	Mangrove plantations, mound plantations, dykes, and embankments; early warning system	Capacity-building through training courses for local government officials in forestry, and organizational change through setting up new functional departments	Coupling of forestry programs to income generation through forest products, fish, and food
Bhutan	Lowering glacial lake levels; deepening river channels; early warning system; climate shelters	Workshops for government officials at the nodal level	Community training in search and rescue, evacuations, and first aid
Cambodia	Climate-proofing of canals and communal ponds; experimentation with crop variation and diversity	Education sessions for provincial and local officials	Local empowerment over prioritization of climate-proofing schemes
Maldives	Sea walls; replenishment of sea ridges; mangrove afforestation; beach nourishment; coral reef propagation; repositioning of water tanks	Decentralization of adaptation planning and management to local political units	Community control over adaptation investments

work best not by improving technology alone, but by seamlessly strengthening three types of adaptation – infrastructural, organizational, and social – to bolster ecosystems, communities, and human organizations.

Such a finding – that resilience and adaptation are interstitial – has been confirmed by a few recent studies. A research team from the World Resources Institute (WRI) investigated 135 case studies of adaptation efforts in developing countries, and noted that a combination of three types of adaptive efforts were most useful:

- Building responsive capacity, such as improving communication between institutions, or enhancing the mapping or weather monitoring capability of a government institution;
- Managing climate risks, such as disaster planning, researching drought resistant crops, or climate-proofing infrastructure;

- Confronting climate change, such as relocating communities or repositioning infrastructure in response to flooding or glacial melting.[98]

Similarly, the WRI, in collaboration with United Nations Development Program, United Nations Environment Program, and World Bank, argued that three dimensions to resilience exist and must be promoted synergistically. Ecological resilience refers to the disturbance an ecosystem can absorb without changing into a different structure or state. Disturbances can be natural, like a storm, or human-induced, such as deforestation. Social resilience refers to the ability of a society to face internal or external crises and still cohere as a community and possess a sense of identity and common purpose. Economic resilience refers to the ability of an economy to recover from shocks, and often entails having a diversified economy composed of members with a variety of different skills.[99]

However, multidimensional resilience also entails risks: degradation or destruction along one dimension of resilience can affect the others; the influence can be both positive and negative. Depleting a forest, for instance, could reduce ecological resilience, which in turn creates fewer jobs (affecting economic resilience) and erodes the community's social resilience (by causing a high proportion of migration or dissention within the community). Conversely, enhanced ecological resilience can improve rents and revenue from logging (economic resilience) and also improve business skills and connection with markets (social resilience). Resilience can be reactive, making the present system resistant to change, or proactive, creating one that is capable of adapting to change.[100] The key challenge for future adaptation efforts will be promoting different types of resilience – infrastructural, institutional, community – that do not trade off against each other, where improving one type is not to the detriment of the others.

Second, one of the striking attributes of the LDCF is its voluntary nature. The international community has mobilized more than $600 million for least developed countries, money they were under no real obligation to raise (other than a moral one, and a mandate from the Conference of the Parties of the UNFCCC), no small sum even if it's insufficient to fully implement all of the adaptation projects in need. And the LDCF has raised these funds relatively quickly in about a decade, and at a scale involving almost four dozen countries and hundreds of partnering institutions.[101] The LDCF does have its troubles related to uncertainty over funding and the complexity of adaptation projects in the context of LDCs, but given that it relies on goodwill it's amazing it works as well as it does.

Third, the LDCF affirms that investments in adaptation pay for themselves quickly, even when being implemented in the poorest countries on the planet. The sorts of adaptation measures being implemented under the LDCF have, according to the Asian Development Bank, a future return on investment as high as 40 to 1. Moreover, these investments in adaptation mitigate carbon dioxide emissions 2 to 500 times more cheaply than alternatives, and they also leverage an additional $4.66 for every $1 committed. The point here is that the costs of inaction clearly outweigh the cost of adaptation.

Fourth, and lastly, this chapter and the LDCF's experience to date demonstrate the value of a functions-based approach to resilience and adaptive capacity rather than an asset-based one. Community or social assets – things like higher wages or better technology – are useless if communities do not have the skills or capacity to use them. Knowledge and assets must be coupled with capacity and improved governance. This creates a more fluid and messy picture of adaptation, but also one that is more realistic. Assets remain only potential until communities leverage them, and adaptation programs must find ways to improve living standards. Viewing adaptation in this way requires conceptualizing resilience not only as infrastructure and technology, but also as the broader social and economic forces that need to occur so that communities can use their assets to manage climate risks.

9
Responsibility and Ecuador's Yasuní-ITT Initiative

Introduction

This chapter argues that environmental protection is a strong energy justice concern, given our duties to nonhuman species as well as the dependence that many communities have on ecosystems for things like fresh air, drinkable water, and edible fish. After detailing how the exploration and drilling of crude oil can devastate ecosystems and communities, the chapter showcases the Yasuní-ITT initiative in Ecuador. That initiative, promoted by President Correa in 2007, would leave almost one billion barrels of crude oil in the ground beneath the Ishpingo Tambococha Tiputini (ITT) oilfield within the country's *Parque Nacional Yasuní*, or Yasuní National Forest. In exchange for not developing the ITT oilfield, President Correa has asked the international community for "fair compensation" of $3.6 billion. The chapter exhibits how prohibiting ITT development brings with it a suite of positive benefits, including the conservation of precious biodiversity, the protection of indigenous peoples, revenues for poverty alleviation and economic development, and the displacement of millions of tons of greenhouse gas emissions.

Responsibility as an energy justice concern

When presenting the case for environmental protection as an energy justice concern, this chapter draws from four particular notions of "responsibility": a responsibility of governments to minimize environmental degradation, a responsibility of industrialized countries involved in emitting greenhouse gases to pay to address the problem of climate change, a responsibility of current generations to protect future ones, and a responsibility of humans to recognize the intrinsic value of nonhuman species.

First, governments (and, indeed, other actors such as corporations and community groups) have a responsibility to minimize the destruction of the environment. As justice theorists Nicholas Low and Brendan Gleeson explain, "Environmental quality is a central aspect of wellbeing for individuals and communities, and it is therefore a critical question for justice."[1] People have a "positive right" to a clean and safe environment and to levels of emissions that provide them with a happy and healthy life, and they also have the "human right not to suffer from the disadvantages generated by global climate change."[2] Far from being a farcical and abstract idea, this distinct notion of responsibility has already found its way into international law. United Nations General Assembly Resolution 45/94 states that "All individuals are entitled to live in an environment adequate for their health and well-being," and 53 nations (by the latest count) have included the provision of such a right in their constitutions. The Draft Declaration of Principles on Human Rights and the Environment, backed by a United Nations delegation, stated in 1994 that all persons should be free from any form of discrimination in regard to actions and decisions that affect the environment, and that all persons have the right to an environment adequate to meet the needs of present generations that do not impede the rights of future generations.[3] (This notion of responsibility is closely aligned with the principle of prudence in Chapter Six and the principle of intergenerational equity in Chapter Eight.)

Second, the primary culprits behind climate change – those countries responsible for most emissions so far – have a duty to take corrective action to try to fix the problem and minimize damage. Essentially, the activities of one group of persons and countries have overused the atmosphere as their carbon "dump," causing injury to a different, much larger group. The situation creates a matter of "corrective" justice, since one group has engaged in wrongfully hurting another group, meaning they should desist from their harmful actions and also compensate them for damages.[4] Polluters and industrializing countries should be required to "clean up their own mess." Part of the argument is historical, since industrializing countries have emitted the most into the atmosphere to reach current levels; part is also an argument from ecology, since one could characterize the atmosphere as a sink with limited space. Rich countries have exhausted the capacity of this sink, denied other countries their shares, and are required to pay compensation for this overuse.[5] Such inequalities must be "reversed," the thinking goes, by imposing extra burdens on those countries and peoples responsible for inflicting and producing atmospheric inequalities.[6] Achieving "justice" means that, when a party has in the past taken unfair advantage of others by imposing costs upon them without their consent, those who have been disadvantaged can demand that the party shoulder burdens sufficient to undo the unfair advantage previously gained – put another way, historical emitters compensate for their misuse of the atmosphere.[7]

To put this type of thinking in perspective, those from developing countries which have contributed few emissions are quick to point out that, if the situation were reversed, the problem of climate change might not even exist. During the discussions for the 2009 Copenhagen Accord, President

Marcus Stephen of Nauru traveled nearly 8,000 miles from his tiny Pacific island to plead for action. He argued that "what if the pollution coming from our island nations was threatening the very existence of the major emitters? What would be the nature of today's debate under those circumstances?"[8]

A third type of responsibility involves future generations. Oxford University moral and political philosopher Brian Barry put it best when he wrote that "from a temporal perspective, no one generation has a better or worse claim than any other to enjoy the earth's resources" and that "the minimal claim of equal opportunity is an equal claim on the earth's natural resources."[9] Essentially, whatever we decide a nice life is, say we collectively value it as a "10," its value as we enjoy it today should be sustained into the future so that future generations do not fall below our level of "10."[10] The present generation cannot claim that it is entitled to a larger share of goods supplied by nature, because most of our technology and capital stock is not the sole creation of the present. We inherit it, and therefore natural resources fall outside any special claims based on the present generation "deserving them." As Barry writes, "since we have received benefits from our predecessors some notion of equity requires us to provide benefit for our successors."[11] This creates a compelling duty to preserve those resources – in this case, a stable climate, unpolluted air, clean water – out of a sense of reciprocity.

Lastly, and perhaps most broadly and radically, we have a responsibility to protect nonhuman species, or at least a responsibility not to destroy their habitats or kill them. Indeed, the claim that we need to adhere to an "environmental ethic" that respects nature for its own sake has a long history, going back at least to the eighteenth-century works of William Blake, William Wordsworth, and Johann Wolfgang von Goethe (among others). These thinkers argued that nature should not be viewed as an impersonal machine, but an organic process with which humanity is united. The transcendentalists in New England referred in similar terms to the sacred dimension of Nature. Henry Thoreau held that Nature was a source of inspiration, vitality, and spiritual renewal, writing that "in wildness is the preservation of the World".[12] The works of Charles Darwin and George Perkins Marsh described how species of plants and animals are part of nature in continuity with other forms of life, including humans, and John Muir circulated the philosophy of wilderness preservation. A few decades later, during the Great Depression and the Dust Bowl in the 1930s, many Americans witnessed perhaps the country's worst ecological disaster at first hand, further inspiring a conservation ethic.

These thinkers and events all catalyzed a modern conception of environmental ethics which strongly argues that humans are dependent on nonhuman forms of nature; pollution of air, water, and land is clearly detrimental to human life; limits should be set on the exploitation and use of natural resources; and humans have a duty to preserve the biosphere for future generations. While certainly supported on spiritual and ethical grounds, the concept of an environmental ethic also has its connections to advances in ecology, conservation biology, and evolutionary biology. These sciences developed the idea that species exist as part of an ecosystem; that humans are interdependent with all members of a biotic community; that biological diversity is needed for ecological balance and stability; and that finite limits exist for population growth and the capacity of the environment to provide resources. As such, we ought to preserve nature for its own inherent sake.

Unfortunately, the modern exploration and production of crude oil contravenes each of these four types of responsibility. It destroys and interrupts nature's ability to provide humanity with ecosystem services; it poisons the atmosphere and contributes to climate change; it robs future generations of their ability to enjoy a good life; and, at times, it mercilessly ravages the world's forests and decimates the biodiversity that depends on them.

Each stage of the crude oil fuel chain – exploration, drilling, refining – poses serious and unavoidable risks such as degraded ecosystems and diminished human health. Exploration necessitates heavy equipment and can be quite invasive, as it involves "discovering" oil and gas deposits found in sedimentary rock through various seismic techniques such as controlled underground explosions, special air guns, and exploratory drilling.[13] The construction of access roads, drilling platforms, and their associated infrastructure frequently induces environmental impacts beyond the immediate effects of land-clearing: they open up remote regions to loggers and wildlife poachers. Every kilometer of new oil and gas roads deforests about 1,000 to 6,000 acres of land, and, in Ecuador, an estimated 2.5 million acres of tropical forest were colonized due to the construction of just 500 kilometers of roads for oil production.[14]

The production and extraction of oil and gas, which are themselves toxic as they contain significant quantities of hydrogen sulfide, a substance that is potentially fatal and extremely corrosive to equipment such as drills and pipelines, is even more hazardous. Drilling for oil and gas involves bringing large quantities of rock fragments, called "cuttings," to the surface, and these cuttings are coated with drilling fluids, called "drilling muds," which operators use to lubricate drill bits and stabilize pressure within oil and gas wells. The quantity of toxic cuttings and mud released for each facility is gargantuan, ranging from 60,000 to 300,000 gallons per day. In addition to cuttings and drilling muds, vast quantities of water contaminated with suspended and dissolved solids are brought to the surface, creating what geologists refer to as "produced water."[15] The average offshore oil and gas platform releases about 400,000 gallons of produced water back into the ocean or sea every day. Produced water contains lead, zinc, mercury, benzene, and toluene, making it highly toxic and often requiring operators to treat it with chemicals, increasing its salinity and making it fatal to many types of plants, before releasing it into the environment. The ratio of waste to extracted oil is staggering: every 1 gallon of oil brought to the surface yields 8 gallons of contaminated water, cuttings, and drilling muds.[16]

The US Geological Survey (USGS) estimates that there are more than 2 million oil and natural gas wells in the continental United States. But the most intense areas of oil and gas production are off the shores of the Gulf of Mexico and along the northern coast of Alaska. Offshore oil and natural gas exploration and production in the Gulf of Mexico exposes aquatic and marine wildlife to chronic, low-level releases of many toxic chemicals through the discharge and seafloor accumulation of drilling muds and cuttings, as well as the continual release of hydrocarbons around production platforms.[17] Drilling operations there have generated massive amounts of polluted water (an average of 180,000 gallons per well every year), releasing toxic metals, including mercury, lead, and cadmium, into the local environment.[18] One study found that mercury levels in the mud and sediments beneath the oil platforms in the Gulf of Mexico were 12 times greater than acceptable levels under federal Environmental Protection Agency (EPA) standards.[19] This exposure to these chronic environmental perturbations continues to threaten marine biodiversity, and human health, over wide areas of the Gulf, to say nothing of when accidents such as *Deepwater Horizon* occur. The Natural Resources Defense Council also noted that the onshore infrastructure required to sustain oil and natural gas processing in the United States has destroyed more coastal wetlands and salt marsh than can be found in the total area stretching from New Jersey through Maine. Similarly, the US Minerals Management Service has documented 70 oil and natural gas spills between 1980 and 1999 from just the production of oil and natural gas. Oil and natural gas spills accounted for more than three million gallons of fuel released into the Gulf of Mexico alone.[20]

The next stage of the fuel cycle, refining, involves boiling, vaporizing, and treating extracted crude oil and gas with solvents to improve their quality. The average US refinery processes 3.8 million gallons of oil per day, and about 11,000 gallons of its product (0.3 percent of production) every day of operation escapes directly into the local environment, where it can contaminate land and pollute water.[21] Moreover, refineries are the second largest source of noxious air emissions in the country, after electricity generation.

As one telling example of how severe the impacts of exploration, production, and refining can become for local communities, consider the case of Azerbaijan, where oil and gas exploration, production, transportation, and refining have been going on since the 1860s (longer than in the United States). Oil production upstream in Azerbaijan is an environmentally devastating activity: large oil fields in the Caspian Sea typically discharge an estimated 100,000 tons of oil directly into water each year. About 30 percent of Azerbaijan's entire coastal area and half of its larger rivers have been contaminated by petroleum development and deemed unsafe for drinking and agriculture.[22] About 90 percent of noxious pollutants emitted into the air in Azerbaijan are connected to the oil industry, and Baku's oil refineries release concentrations of hydrocarbons, sulfur dioxide, and nitrogen oxides at levels considered unsafe by international standards. Epidemiological studies have confirmed that an abnormally high percentage of children near oil-producing facilities in Azerbaijan are born premature, stillborn, or with genetic diseases – born into families living in degraded landscapes like the one in Figure 9.1.[23] To put the findings from these studies in perspective, average life expectancy for someone living near these Azerbaijani oil fields is less than 40 years compared with a global average above 70 years. These statistics may explain how the country earned the prestigious title of "worst polluted place in the world."[24]

In the northern parts of the Caspian Sea, near Kazakhstan, environmental damage from oil production is so severe that the United Nations calculates it exceeds the profits generated by oil development. One dramatic accident occurred in 1985 when an oil well near Tengiz caught fire and shot a 200-meter column of flame into the air that took more than one *year* to extinguish, burning 3.5 million tons of oil, producing 500,000 tons of hydrogen sulfide, and negatively impacting biodiversity and human health within a 100-kilometer radius. Six similar accidents have occurred since among offshore oil platforms in Turkmenistan waters.[25]

Oil and gas development in the western Amazon in South America, the most biologically rich part of the Amazonian rainforest, has already accelerated deforestation and caused widespread contamination through accidents and spills. The newly completed Camisea pipeline began operations in 2004 and has already had five major spills, to say nothing of oil spills

Figure 9.1   Oil and gas production near Baku, Azerbaijan

at refineries across Ecuador and Peru.[26] In the Oriente region of Ecuador, Texaco's oil-producing facilities have polluted local water tables and accelerated soil erosion.[27]

Nor are the impacts of oil limited to the upstream part of its development. Downstream, as oil and refined petroleum products enter our automobiles, furnaces, and boilers, their combustion releases a variety of unhealthy pollutants and particles directly into the air, where they become ingested and inhaled by human beings and ecosystems, contributing to hospital admissions, acid rain, and ozone depletion. Conventional automobiles are often the largest single human-caused source of particulate matter (PM), and, for those places with stringent emissions requirements for vehicles, such as California or the EU, the second largest human source after power plants (forest fires and dust storms are the leading nonhuman sources). Thousands of medical studies have strongly associated inhalation of PM with heart disease, chronic lung disease, and some forms of cancer. Their findings were confirmed, yet again, when the burning of 650 oil wells in Kuwait during the 1991 Persian Gulf War affected US soldiers with respiratory illness, asthma, and emphysema.

In the United States, deaths from PM pollution are comparable to those from Alzheimer's disease and influenza, and greater than the deaths from nephritis, septicemia, breast cancer, automobile accidents, prostate cancer, HIV-AIDS, and drunk driving.[28] In France, the Agency for Health and Environmental Safety projects that normal automobile emissions kill 9,513 people per year and result in 6–11 percent of all lung cancer cases identified in people above 30 years of age.[29] Another report from the World Health Organization investigating pollution in Austria, Switzerland, and France calculated 40,000 deaths per year due to PM emissions from automobiles.[30]

Case study: Ecuador's Yasuní-ITT Initiative

Perhaps the most distinguishing aspect of Ecuador's Yasuní-ITT initiative is that it averts *all* of the externalities and environmental insults associated with almost a billion barrels of oil by keeping it underground. Yasuní, both a national park and an internationally accredited biosphere reserve in the eastern part of Ecuador, houses an old-growth rainforest in the Amazon and a number of distinct indigenous cultures. The Yasuní-ITT Initiative proposes that the international community donates to Ecuador $3.6 billion – roughly half the value of the oil found there – in exchange for not developing the Ishpingo Tambococha Tiputini (ITT) oilfields. The proposal places the $3.6 billion into social and environmental development programs, including "massive programs of reforestation and natural forest recovery"[31] and the promotion of domestic renewable energy. The proposal strikes a balance between protecting the park and its indigenous people, and creating revenue for the Ecuadorian state, already heavily dependent on oil taxes, a major oil exporter, and a member of OPEC. The proposal also will sink – that is, store and keep out of the atmosphere – more than 400 million tons of carbon dioxide. The initiative is therefore an innovative way to address global warming, protect biodiversity, support indigenous peoples, and promote social development.[32] Because of its ability to capture climatic, environmental, and social goals simultaneously, independent assessments have called it a "visionary proposal,"[33] "striking in its scope and its creativity,"[34] and a "precedent-setting advance towards avoiding oil and gas development in sensitive areas of mega-diverse developing countries."[35]

History

The Yasuní National Forest (YNF), known locally as the *Parque Nacional Yasuní*, sits about 190 miles (306 kilometers) east from Ecuador's capital city of Quito along the border of Peru in the Amazon River Basin. The YNF lies almost directly on the equator and encompasses several rivers that feed into the Amazon, and it expansively covers 2.4 million acres (971,246 hectares) of low-lying, flat plains dotted with foothills, 1.8 million (728,424 hectares) of which constitute a "no-go" or "intangible" zone, known locally as the *zona intangible*. The government created the YNF in 1979 and the United Nations Educational, Scientific and Cultural Organization declared it a World Biosphere Reserve in 1989.[36] Figure 9.2 shows its general location within South America.[37] As Figure 9.3 shows, the ITT oilfield rests beneath the eastern parts of the forest. Geologists estimate that 846 million to 1.5 billion barrels lie within these reserves, accounting for 20 to 30 percent of the country's entire reserves, with most projections agreeing on the presence of 900 million barrels, making it Ecuador's second largest oil reserve, after those feeding Shushufindi and Auca.[38] In order to evacuate such oil out of the YNF, it would have to be transported via rail and pipeline to Ecuador's coast and then refined.[39]

The idea of protecting the ITT oilfields from exploitation started in the late 1990s. In 1997, civil society groups such as Acción Ecológica and Oil Watch put together a proposal known as "Option One," which considered leaving the oil "permanently underground" in exchange for "compensatory financial mechanisms."[40] This proposal, which ended up gaining the support of more civil society groups and a collection of scientists, was intensely discussed within the Ecuadorian government but was never acted upon, since the Brazilian company Petrobras had purchased rights to the ITT field and had begun the construction of a road in 2005 so it could enter the YNF to commence oil production. In April of that year, however, President Lucio Gutierrez was deposed after widespread social protests over corruption in the Ecuadorian courts, and in July the new government, riding a wave of reform, revoked Petrobras's permit for the road. Two years later, in March 2007, President Rafael Vicente Correa announced his support for "Option

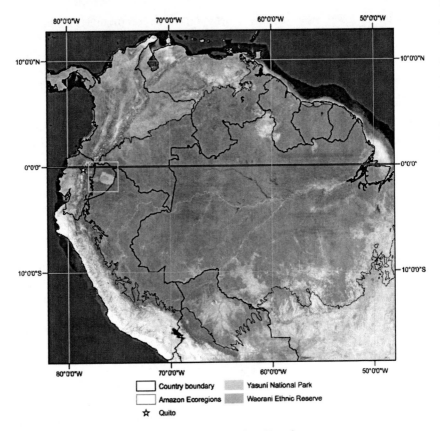

Figure 9.2   Location of the Yasuní National Park in Ecuador
Source: Adapted from Finer *et al.*

One," publicly renamed the "Yasuní-ITT" initiative. During his announcement, President Correa commented that Yasuní-ITT was "an extraordinary example of global collective action, that would allow not only reduced global warming, which benefits the whole planet, but also introduce a new economic logic for the 21st century, which assigns a value to things other than merchandise."[41]

More specifically, Correa's proposal consisted of four parts:

- not extracting crude oil from any part of the ITT oil field;
- channeling international resources from other countries and corporations into a sort of ecosystems payment scheme to compensate Ecuadorians for refraining from oil development;
- creating a capitalization fund whose interest would be utilized to provide a permanent source of income for those living within the YNF;

- providing Ecuador with the resources to eventually transition to a "post-petrol" economy.

In terms of compensation, Ecuador's state petroleum company, PetroEcuador, estimated at that time that the ITT's reserves would bring in roughly $720 million in investment per year at a price of $80 per barrel. The government thought it fair to ask for slightly more than half that, $360 million, over ten years, giving a total of $3.6 billion, and after some internal debate the President set the target of receiving $100 million by December 31, 2011, rising eventually to the entire $3.6 billion by 2024.[42] In asking for $3.6 billion, the President was undoubtedly being generous. He was not only forgoing the true value of the ITT field ($7 billion) but precluding infrastructure and roads that would enable logging and the harvesting of the forest, additional sources of revenue for the government. As one independent study put it, "the total value to Ecuador of pursuing the Yasuní-ITT Initiative is considerably higher" than what the President asked for.[43]

After President Correa reaffirmed his commitment to the Yasuní-ITT initiative in front of the United Nations General Assembly later in 2007,

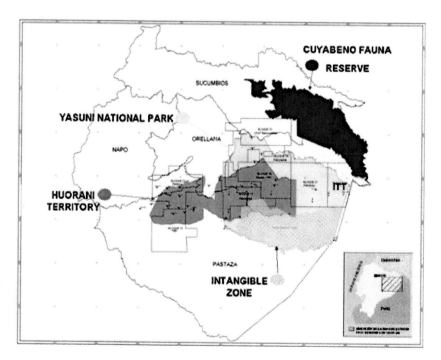

Figure 9.3 **Yasuní National Forest and the ITT Oil Fields**
Source: Adapted from Finer *et al.*

the Ecuadorian government issued Decrees 847 and 882 in January 2008, authorizing the establishment of a "National Development Fund" and creating a Technical Secretariat. These decrees indicated that payments into the initiative could take the form of cash, or as credits against or a reduction of Ecuador's foreign debt. The National Development Fund was to "help the country promote and develop renewable energy projects, transportation systems, programs to eliminate poverty, and equitable access to health care and education."[44] That same year, Petrobras, which technically retained the rights to Block 31 (handed over before the proposal was made in 2007), announced that it had left the country and returned the reserves to the government.

The proposal changed slightly during the middle of 2008, partly due to the global economic crisis and difficulties in raising funds towards the $100 million target (a deadline set for December 31, 2011). In August 2008, Ecuador declared that, rather than have donors purchase the oil, they would issue "Yasuní Guarantee Certificates," or YGCs, for the carbon dioxide locked within the ITT oilfields, with the idea that YGCs would become fungible commodities on the international carbon credits market.[45] YGCs are treated as "noninterest bearing fiduciary papers issued as a guarantee that the ITT oil will stay underground forever," meaning that, if Ecuador violates the proposal and starts producing oil, certificate holders can "redeem their investment."[46]

The 2008 revisions also stipulated that income from donations and the sale of certificates would enter a "Yasuní ITT Trust Fund" used to finance "sustainable development projects" throughout Ecuador. That Fund, formally created in August 2010, exists to "receive and manage contributions in support of Ecuador's historic decision to forego extracting oil." It is currently administered by the Multi-Partner Trust Fund (MPTF) Office of the UNDP, and is managed by a Steering Committee consisting of three government members, two representatives of contributing countries, two members of UNDP, and one member of civil society. Ecuador chose the UNDP to supervise its fund because the MPTF Office already manages $5.5 billion in annual contributions received from more than 70 development partners and 80 countries entering 40 separate accounts.[47]

As of December 2011, the Yasuní-ITT initiative squeaked past its goal of $100 million and had raised precisely $116,848,384.[48] It has been formally supported by a number of prestigious figures, including the Nobel Laureates Muhammad Yunus and Desmond Tutu (among others), Ban Ki-Moon (the Secretary General of the United Nations), Prince Charles of Great Britain, and ex-presidents Mikhail Gorbachev from Russia, Felipe Gonzalez from Spain, Fernando Henrique Cardoso from Brazil, and Ricardo Lagos from Chile. It has received official backing and financial contributions from the German Parliament and the governments of France, Hungary, Italy, and Spain, as well as international institutions such as the European Union, OPEC, the Andean Community of Nations (CAN), the Andean Development Corporation

(CAF), the Organization of American States, the South American Union of Nations (UNSAR), and the Inter-American Development Bank (IADB). Prominent civil society organizations have declared their support, including the International Union for Conservation of Nature and Natural Resources (IUCN) and dozens of local groups and nongovernmental organizations in Ecuador.[49] The Initiative was previously open to only governments and corporations pledging more than $100,000, but, in an effort to attract more donors, individuals and businesses wishing to contribute amounts as small as $25 can now participate.[50] Momentum is needed, as the Ecuadorian government has established a new target of securing $291 million in contributions in both 2012 and 2013 to keep the initiative going. In tandem with this new goal, the Yasuní-ITT Secretariat is expanding its fundraising efforts to Australia, Europe, and North America and starting to use social networking tools such as Facebook and Twitter to accelerate outreach.

Benefits

The Yasuní-ITT initiative has five distinct benefits: (1) the conservation of biodiversity, (2) the protection of indigenous peoples, (3) economic and social development, (4) displacement of greenhouse gas emissions, and (5) a surprisingly positive cost curve exceeding the traditional value of the crude oil.

Conservation of biodiversity

Probably the most significant benefit is the preservation and protection of precious biodiversity found nowhere else on the planet. Scientists note that the YNF "contains extraordinary biological and cultural richness, along with abundant natural resources" and "is one of the most bio-diverse places on Earth, and the core of a unique area where the continent's plant, amphibian, bird, and mammal species richness centers overlap."[51] Scientists have praised the YNF for having the "highest level of biodiversity on the planet,"[52] dubbing it the "cradle of the Amazon" and the "most biologically diverse hotspot in the Western Hemisphere."[53] The World Wildlife Fund declared the Humid Napo Forest within the reserve to be the first in priority on a list of 200 "most important" biodiversity areas on the planet. (Ironically, it is the past death of much of this biodiversity that makes the region home to so much fossilized carbon, as oil is merely long-decayed organic matter.)

The YNF currently shelters more documented insect species than any other forest in the world, and it is among the most diverse forests in the world for different species of birds, bats, amphibians, epiphytes, and lianas. Scientists have spent the past five decades trying to catalogue the species that roam or live within the YNF, and some of their statistics are startling. The Tiputini Biodiversity Station, located adjacent to the northern border of the park and established in 1994, holds world records for local amphibian, reptile, and bat species richness. The Yasuní Research Station, also established in 1994, holds the world record for number of tree species found in one hectare (2.5

acres) of land, with 655 – meaning it has more trees in a single hectare of land than the total number recorded in all of Canada and the United States combined. As Table 9.1 reveals, the YNF's concentration of shrub and tree species is also the highest in the world among all existing national parks. Some 4,000 plant species and 2,274 tree species, 499 species of fish, 173 species of mammals, 150 species of amphibians, 121 species of reptiles, 80 species of bats and 610 species of birds live within 10,000 square kilometers of the YNF. The number of insect species is estimated to be more than 100,000 per hectare, the highest concentration on the planet. It is a critical habitat to 28 globally threatened or near-threatened vertebrates, including the giant otter, the Amazonian manatee, the pink river dolphin, the giant anteater, the Amazonian tapir, and the white-bellied spider monkey. The YNF is home to 95 threatened or near-threatened plant species, including eight that are "globally endangered" or "critically endangered." The YNF contains at least 43 species of amphibians, birds, and mammals endemic to the park – that is, found nowhere else on earth.[54] As biologists have exclaimed when trying to count species within Yasuní, "it isn't even diversity any longer – it's hyperdiversity, or megadiversity."[55]

At least four factors appear to explain why the YNF is so diverse and deserving of protection. One is topography and climate: the YNF sits at the

Table 9.1 Global comparison of shrub and tree species richness

Site	Country	Small tree species per hectare	Large tree species per hectare	Rank
Yasuní National Park	Ecuador	655	251	1
Lambir Hills National Park	Malaysia	618	247	2
Pasoh Forest Reserve	Malaysia	495	206	3
Khao Chong Wildlife Refuge	Thailand	380	–	4
Yunnan Province (Xishuangbanna)	China	292	–	5
Bukit Timah Nature Reserve	Singapore	276	113	6
Korup National Park	Cameroon	236	87	7
Palanan Wilderness Area	Philippines	197	100	8
Barro Colorado Island	Panama	169	91	9
Okapi Faunal Reserve (Ituri)	D.R. of Congo	161	57	10
La Planada Nature Reserve	Colombia	154	88	11
Sinharaja World Heritage Site	Sri Lanka	142	72	12
Doi Inthanon National Park	Thailand	105	67	13
Ken-Ting National Park	Taiwan	104	61	14
Huai Kha Khaeng W. Sanctuary	Thailand	96	65	15
Luquillo Experimental Forest	Puerto Rico	73	42	16
Northern Taiwan (Fushan)	Taiwan	34	–	17
Mudumalai Wildlife Sanctuary	India	24.7	19.8	19

Source: Bass *et al.*

intersection of the Andes and the Amazon, which means it has high amounts of rainfall but also a relatively "aseasonal" climate, making it perpetually wet and warm. Another is its large geographic size and wilderness character, enabling it to support large populations of "megafauna" such as the jaguar and the harpy eagle, which cannot survive in smaller parks or reserves. As of 2010, for example, it was still possible to walk continuously through mega-diverse forest from eastern Ecuador to southern Venezuela – a distance of 2,000 kilometers – without crossing a single road. A third factor is historical: the YNF served as a refuge in the Pleistocene era, when glaciers drastically cooled the planet and converted many forests into grasslands, forcing many species, or "Pleistocene refugees," to concentrate in safe havens such as Yasuní. Figure 9.4, for example, shows how more species are concentrated

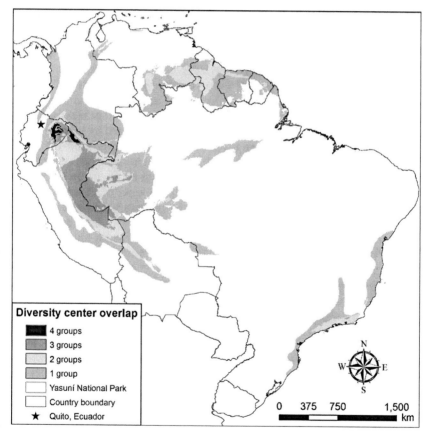

Figure 9.4 Species richness in South America for amphibians, birds, mammals, and vascular plants

Source: Bass *et al.*

within parts of the YNF than any other part of South America.[56] A fourth factor is that Yasuní, unlike many other national reserves, will maintain much of its heritage if left undisturbed. The forest will likely preserve its wet, rainforest conditions in the face of climate change and drought that will affect other national parks around the world.[57]

The conservation value of the Yasuní-ITT initiative is so great because oil development undoubtedly would significantly degrade the forest. Four things make oil development in the YNF especially troublesome. First, a considerable portion of the ITT oilfield lies at the extreme southern end of the block, within the *zona intangible*, some of the richest parts of the park in terms of biodiversity and cultural heritage. Second, however, is that the most profitable reserves are spread almost evenly across the block, meaning development would require six separate drilling platforms and one reinjection platform connected by an extensive access train network.[58] Large tracts of forest would need to be cleared, and everything from drilling equipment and roads to support facilities and pipelines would have to be brought into the interior. Third, apart from the initial impacts of deforestation and construction, the risk of spills and pipeline failures is significant. PetroEcuador currently holds the "world record" for petroleum spills at roughly 400 a year (100,000 million liters of petroleum in total), or more than one per day. Not all of these are accidental, either, with Repsol YPF and other companies reporting that many spills are "on purpose" and instigated by local communities to extort money in exchange for "protection."[59] Fourth, the oil found within the YNF, and indeed the entire Oriente Basin underlying the Amazon, is of comparatively "low" quality and "heavy," with an American Petroleum Institute value of 14.7, meaning it has high amounts of carbon dioxide, chemical pollutants, sulfur, and particulate matter, creating a need for intensive refining and higher emission rates downstream.[60] Indeed, an internal assessment of ITT development conducted by PetroEcuador found that the risk of "direct impacts" from "immediate deforestation" and "project-related spills, leaks, or accidents" would be "substantial."[61] A second, more critical evaluation suggested that the ITT oilfield would need at least 130 wells each producing 65,000 cubic meters of solid waste, equivalent to 13,000 dump trucks' worth being discarded into the forest each year. It also estimated that 390,000 cubic meters of liquid wastes, in addition to a total production of 8.7 billion barrels of produced water, would have to be reinjected into wells and released into the YNF's rivers.[62]

Protection of indigenous peoples

The Yasuní-ITT proposal not only protects biodiversity; it also preserves the hunting grounds and ancestral lands of indigenous peoples. The YNF is the home of the Huaorani and Quechua people, as well as two other indigenous Peoples in Voluntary Isolation, the Tagaeri and the Taromenane. These four groups together consist of about 9,800 individuals spread throughout the park, dependent upon hunting and gathering to subsist. The Huaorani reside on a 612,000-hectare reserve created by Ecuador's government in 1983 in the northern part of the park; the Quechua, descendants of the Incas, reside in the northwest part of the park; the self-sufficient Tagaeri and Taromenane tribes inhabit the medium and lower zone and have remained "voluntarily isolated" and "independent from other civilizations." The subsistence of these groups is already in a precarious state, with activities such as illegal logging, tourism, scientific research, and evangelical missions encroaching on their territory and culture. However, the Tagaeri and Taromenane maintain no peaceful contact with the outside world and are renowned for spearing unwanted intruders, including the murder of a few dozen oil company workers. Numerous studies have confirmed that development of the ITT oilfield would directly threaten the cultural vitality of these groups, their kinship networks, and their ability to remain self-sufficient.[63]

Economic and social development

Even though it relinquishes billions of dollars in potential oil revenues, the Yasuní-ITT initiative does have substantial economic and social benefits. Oil revenues would not directly enrich multinational oil and gas companies; instead, they accrue in a Yasuní ITT Trust Fund that finances "sustainable development" projects throughout Ecuador.[64] The Fund's capital is invested in diversifying domestic energy production towards renewable sources of energy, and the interest on the fund is devoted to preserving the YNF and meeting the goals enshrined by Ecuador's National Development Plan. These include:

- preventing deforestation and managing Ecuador's network of protected areas spread across 4.8 million hectares, or 38 percent of the country's land;
- a reforestation campaign including the replanting of one million hectares of trees among small landowners;
- implementing energy efficiency programs;
- researching and developing renewable energy technologies;
- funding national education, training, health, tourism, and other forestry programs.[65]

The money from Yasuní-ITT will certainly help contribute towards the alleviation of poverty, no small task given that roughly 49 percent of the population lived below the international poverty line in 2006, and that per capita incomes have grown at only 0.7 percent per year from 1982 to 2007.

Indirectly, refraining from ITT oil development could prevent land and resources falling into the hands of the country's drug cartels. Petroleum exploration involves some of the same infrastructural requirements that can process coca leaves into coca paste and into cocaine. For instance, white gas, sulfuric acid, and other chemicals used by petroleum companies are utilized to manufacture illegal narcotics. In this way, developing ITT could open up roads and distribute materials that would invariably increase cocaine production and the trafficking of narcotics.[66]

Displacement of greenhouse gas emissions

Sequestering the crude oil found within the ITT fields keeps between 407 million and 436 million tons of carbon dioxide "sunk" in the ground rather than emitted into the atmosphere. This is no small feat, given that Ecuador emits a mere 29 million tons of per year, while Brazil's annual emissions total about 332 million tons and France's total about 373 million tons.[67] This calculation of ~420 million tons also ignores the "double" benefit of leaving intact millions of acres of old-growth forest which act as a carbon sink.[68]

Positive cost curve

Perhaps oddly, the Yasuní-ITT initiative has a positive cost curve; that is, it produces benefits that exceed the traditional market value of the crude oil found within the ITT oilfields. Presuming those oilfields contain 410 tons of carbon dioxide, their value on the European carbon market would amount to $8.2 billion at a cost of $20 per ton. Additionally, the prevention of deforestation and the reforestation program envisioned by the Trust Fund will save an estimated 859 million tons of carbon dioxide, 791 million tons from prevented deforestation over 30 years, and 68 tons sunk from afforestation projects.[69] Furthermore, the investments expected to be made in climate change mitigation, mostly the expansion of Ecuador's hydroelectric capacity and new investments in energy efficiency, solar, wind, and geothermal energy, will likely displace a further 43 million tons of carbon dioxide; attributing 30 percent of these gains to the ITT initiative amounts to 13 million tons.

Adding the saved emissions from the displaced oil, deforestation, reforestation, and renewable energy together results in 1.282 billion tons of carbon dioxide displaced, worth more than $25.6 billion at a price of $20 per ton. Keep in mind that this estimation is incredibly conservative, given that it looks only at the price of carbon dioxide; it excludes other major benefits to the proposal, including keeping complete the YNF's ability to provide food, medicine, and clean drinking water in addition to other ecosystem services such as flood and drought control and the regulation of rainfall. One study from the Earth Economics Institute hinted at these other benefits when it projected a net present value of $9.89 billion for the "environmental benefits" of the proposal, such as stopping deforestation, loss of ecotourism, and the non-timber-related services of the YNF.[70] If accurate, this brings the total value of the initiative to a mammoth $32.8 billion – more than four and a half times the $7.2 billion at which PetroEcuador valued the ITT oilfields.

Challenges

That said, Yasuní-ITT does face some serious challenges related to (1) finding donors and financing, (2) political pressure to develop the ITT reserves, (3) continued oil development elsewhere in Ecuador, and (4) the risk of leakage through carbon offsets.

Finding donors and financing

The simplest – yet most severe – challenge has been trying to raise money for the proposal. Official project documents expected countries such as Japan and the United States to contribute $1 to $3 billion each, with other major contributions from more than a dozen European countries.[71] However, the first four months of the scheme raised a meager $100,000, a paltry 1/1000th of the first year's requirement.[72] In early 2012, out of the formal target of $3.6 billion, the scheme had so far raised a trifling $116 million, about 3.2 percent of the total. Raising even this amount was arduous, with many donor countries declining to support the initiative for fear that the "avoided emissions" strategy could "lead to similar plans being considered as part of future global warming/climate change treaty negotiations."[73] This concern almost "scared off" Germany from making its contribution, the largest so far, with one high-ranking official stating that "a direct payment into a fund of this type would set a precedent that could ultimately prove very costly."[74] The goal of $291 million by the end of 2012 requires that more than $797,000 be raised *every day* – and the proposal hopes to sustain that amount for the next 11 years, making its longevity and effectiveness far from guaranteed.

Political pressure

A second significant challenge has been understandable political pressure domestically to develop the ITT oil reserves to create much-needed revenue for social programs – revenue that some perceive to be more constant, and immediate, than the "distant" and "uncertain" promises of countries contributing towards the Trust Fund. Political maneuvering was most intense in 2007 and 2008, during the nascent stages of the proposal, when high-ranking government officials publicly opposed the initiative and/or resigned. Alberto Acosta, the Minister of Mines and Energy, responsible for trying to win over support for the project in political circles, noted that "few officials understood the rationality" underlying the proposal and that "for many of my colleagues, it was inconceivable that the very minister in charge of developing oil would campaign for its non-exploitation."[75] His and the President's efforts were publicly disputed by the president of PetroEcuador, who continued to negotiate arrangements with oil companies for the extractive development of ITT in 2007 despite the President's announcement. Other disagreements saw the consecutive turnover of three Foreign Affairs ministers and three entire negotiating teams, with President Correa's public rebuke of his negotiating team at the Copenhagen Climate Summit, where the Trust Fund was to be signed, leading to the resignation of that entire team as well as the President's close friend and Foreign Minister Fander Falconi.

For a nation where oil is the lifeblood of the economy – from 2004 to 2010 petroleum accounted for 57 percent of total export revenue, represented about one-quarter of national GDP, provided 47 percent of the country's energy, and contributed to about one-third of the government's annual budget – some have called the loss of $7.2 billion worth of oil "too painful to be fiscally responsible."[76] The UNDP admits that "preserving Yasuní will require a monumental sacrifice from the Ecuadorian people, a third of whom live below the poverty line."[77] Ecuador could certainly use the money today rather than tomorrow, with PetroEcuador reporting a 15 percent decrease in overall oil production and government statistics revealing a 6.9 percent rise in employment, a 51.7 percent rate of sub-employment (workers who consider themselves underpaid or underemployed), a 12 percent decrease in overseas remittances due to economic crises affecting the United States and Spain, and a 0.4 percent contraction of GDP in 2009. As Oxford academic Laura Rival notes, "the current government has found it difficult to resist the short term incentive of continuing to expand oil production, and this despite its full awareness of the very low quality of the oil being currently extracted, the proven limits of existing reserves (Ecuador will cease to be an oil producer in the next three decades), and the very serious ecological damage being caused by the expansion of the oil frontier."[78]

When mentioning this pressure, however, it must be stated that the public as a whole has been generally supportive of the scheme. One editorial from an Ecuadorian writing in the province of Orellana stated that:

> The only dignified thing to do is to leave the oil in the ground. This in no way will make us poorer. The country generates sufficient revenues today for us to start organizing things differently, and realize our dreams. The president knows this very well.[79]

After President Correa announced the initiative in 2007 at the United Nations, polling indicated that 78 percent of Ecuadorian citizens supported it.[80] Another poll near the end of 2011 indicated that almost 90 percent of the public approved of it.[81]

Nonetheless, given both political pressure and strong economic incentives to generate revenue, Yasuní's protection could be temporary. If the initiative is unable to raise the needed capital from international donors, the country has repeatedly indicated it will have no choice but to develop the ITT oilfield. Indeed, Ecuador's own Ministry of Energy and Mines has revealed that it is still considering several options for the ITT oilfields, only one of which leaves it unexploited.[82] Legally, oil drilling within the YNF can proceed if the

Ecuadorian Congress declares it to be "in the national interest" and then the President approves it.[83]

Interestingly, and positively, the Yasuní-ITT proposal does have a built-in safeguard to prevent this happening. In exchange for giving contributions for YGCs, if the Ecuadorian government does decide to drill within the reserve it will have to give all of the money raised back to donors. Every YGC issued above $50,000 comes with a guarantee that "in the unlikely scenario that in the future, the government of Ecuador decides to extract oil from the ITT field, the YGCs will be made redeemable and the State will lose ownership of the raised funds, which will be transferred back to the contributors."[84]

Continued oil development elsewhere

A third challenge is that the Yasuní-ITT proposal does not stop oil production, and its pernicious environmental and social consequences, in other parts of Ecuador, even on the periphery of the YNF. Oil production has a long history in Ecuador, dating back before colonization, when indigenous peoples used it to light torches and the Peninsula of Santa Elena, near Guayaquil, had natural oil drains used by the Spanish to waterproof their boats. Modern oil exploration dates back to 1878, when the National Assembly granted rights for drilling and the five decades spanning the 1930s to 1970s saw intensive oil exploration and drilling by companies such as Texaco-Gulf (now Chevron), Shell Oil, and the Ecuadorian State Oil Corporation to the point where Ecuador joined OPEC in 1974.[85] Today, as shown in Table 9.2, Ecuador

Table 9.2   Oil production by companies in Ecuador, 2007

Company name	%
Petroproduccion	36.55
Sipec	3.59
PetroEcuador-Perenco	1.11
PetroEcuador-Occidental	5.34
Petroleos Sudamericanos	1.23
Tecpecuador	1.59
Petrobell	0.87
Perenco	3.05
Occidental	13.50
Encana Ecuador	1.62
Repsol YPF	9.91
AEC Ecuador	10.24
City Oriente	0.74
AGIP Oil Ecuador	5.41
Ecuador TLC	4.85
Espol	0.38
Canada Grande	0.03

Source: Adapted from Pelaez-Samaniego *et al.*

has 16 companies vying for the country's five billion barrels of oil reserves. It has more than doubled its oil production from 220,000 barrels per day in 1990 to 510,000 barrels per day, two-thirds of which is exported to North American, Asian, and Latin American markets. 2010 even saw the government attempting to boost oil production by investing $700 million in oil infrastructure.[86]

Much of this oil production occurs within protected areas, and it is encroaching upon the preserved land of the YNF. Existing leased oil blocks almost entirely surround the YNF, and Figure 9.5 shows how close some of those blocks are to protected areas of the park. The Yasuní proposal, while innovative, does nothing to prevent the pollution, deforestation, and destruction of habitats in these areas, nor does it address the displacement of indigenous peoples throughout these leased blocks. One environmental group has accused Chevron of undertaking "reckless oil development"

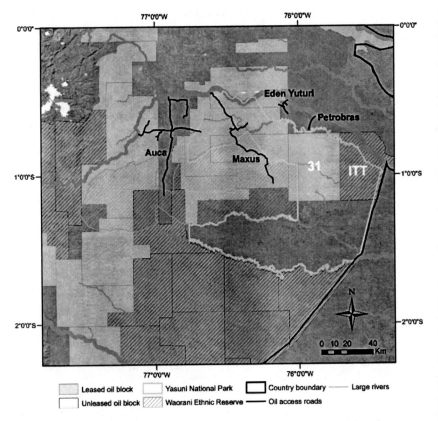

Figure 9.5   Oil blocks and oil access roads within the Yasuní region of Ecuador

Source: Adapted from Bass *et al.* and Finer *et al.*

throughout the area that has "taken a devastating toll on Ecuador's northern Amazon forest communities and ecosystems" and its historical "dump and run tactics have left the rainforest floor stained with toxic waste pits and streams laced with heavy metals and known carcinogens."[87] Oil development has played a "major role" in turning the Napo region, once known for its biodiversity, into one of the 14 major deforestation fronts in the world, and it has contributed to Ecuador having one of the highest deforestation rates in Latin America for several years. Similarly, another site known as Santa Cecilia, located just to the north of Yasuní, once had some of the largest concentrations of amphibian and reptile diversity in the world, but is now "completely deforested due to oil-related disturbance and colonization."[88] In the five years since President Correa touted the Yasuní-ITT proposal, oil development has continued to spread across eastern Ecuador and, across the border, in southwest Columbia and northern Peru. The scientists on the ground at the Tiputini Biodiversity Station within the YNF remark that it's almost as if an "unstoppable force is closing in" and that "you can hear the generator of an oil platform from a few kilometers' distance, and at night you can see the glow of the gas flame."[89]

Leakage and carbon offsets

A final challenge to the initiative involves its dependence on global carbon markets as primary financing mechanisms for the Yasuní Trust Fund. The core of this challenge relates to "offsetting," the fact that YGCs can be used to "offset" the emissions of carbon dioxide and greenhouse gases in other countries. Put another way, a country wishing to emit a ton of carbon dioxide can still do so as long as it "offsets" that ton by purchasing a carbon credit or conducting an activity somewhere else to reduce greenhouse gas emissions, such as buying YGCs. Thus, for every single ton of greenhouse gas avoided or offset abroad by the Yasuní-ITT initiative, an investor is permitted to emit one more ton at home, meaning the practice of offsetting does not encourage a reduction in total greenhouse gas emissions, only that countries offset their emissions elsewhere, making them carbon-neutral. In this way, in the short term at least, by acting as offsets YGCs do little to truly address the potentially irreversible impacts from climate change. A related concern is that, by involving such a large number of carbon credits – more than 400 million related to the oil alone, potentially hundreds of millions more for the carbon sunk in the biomass within the YNF – the scheme will deflate their value by contributing to an oversupply.[90]

Conclusion: lessons and implications

The Yasuní-ITT initiative raises at least five salient conclusions regarding energy security, climate policy, and energy justice. First, as with many of the other case studies throughout the book, the proposal required a political champion with the vision and strength to conceive, implement, and defend a controversial plan to leave billions of dollars' worth of oil unexploited. At a price of $180 per barrel – a possibility within the next decade – the ITT oilfield would be worth a staggering $18.4 billion. Granted, the analysis conducted in this chapter suggests that the benefits of leaving the forest intact and the oil underground could be even greater ($32.8 billion), but President Correa needed courage and foresight to support the proposal. For a nation deeply dependent on oil exports, on oil for its domestic energy consumption, and on oil for its government revenue, it is surprising that the proposal existed in the first place, and striking that it has been implemented.

Second, and akin to a number of the book's other case studies, is the polycentric nature of the proposal, which sees actors as small as individuals and self-organized civil society groups working with the government of Ecuador to make contributions to the Yasuní-ITT Trust Fund. That Fund is managed by an intergovernmental organization, the UNDP, and is currently being replenished by a collection of European governments and regional institutions.

Third, though the Yasuní-ITT initiative is the first of its kind in the world, it represents a revolutionary mechanism that other countries wishing to protect natural resources or indigenous communities can point to. It has, most tellingly, succeeded where other programs have faltered. Yasuní-ITT goes far beyond the limited results of the Kyoto Protocol and some of the inequalities inherent in its Clean Development Mechanism, essentially keeping an amount of emissions greater than France's from entering the atmosphere (if it is fully implemented). It was able to protect the YNF, at least for now, far faster and more effectively than the United Nations Reducing Emissions from Deforestation and Forest Degradation (REDD) program in developing countries, which, as of July 2012, has raised $117 million (roughly the same as Yasuní-ITT), but had to spread that amount across 46 partner countries, rather than investing it solely in Ecuador.[91] Yasuní-ITT was able to raise funds without requiring debt-for-nature swaps that necessitate groups like the World Bank and International Monetary Fund forgiving large amounts of debt. The efficacy of Yasuní compared with the Clean Development Mechanism, REDD, and debt-for-nature swaps suggests that a fundamental reexamination of their assumptions may be called for. It also strongly implies that a better path exists for planners and policymakers trying to balance economic development with environmental and social sustainability.

Fourth, however, is that there may be limits to the number of Yasuní-ITT-like proposals the international community will support. The proposal brings to light a critical but uncomfortable question concerning climate change and the need to keep carbon in the ground: who pays for stranded assets? As one colleague recently remarked at a conference on energy governance, "I'm not sure we will see more initiatives like Yasuní-ITT – did horse and carriage owners get or deserve compensation when the automobile was invented, and they found their technologies displaced?"[92] There is an element of regressiveness to the proposal, in paying countries and communities to do things they should already be doing on their own. Given the current malaise plaguing the global economy, with rates of historically low growth and high unemployment, finding the billions (and perhaps trillions) of dollars needed to keep all the needed carbon stored in existing oil, gas, and coal reserves in the ground may be a nonstarter.

Fifth, and perhaps most provocatively, the Yasuní-ITT proposal turns upside down how the current global market values oil and subverts fundamental economic principles regarding the pricing of commodities. Yasuní-ITT completely inverts the value of fossil fuels and petroleum from something that ought to be explored, drilled, produced, and combusted to something best left where it is to sequester carbon and preserve biodiversity. President Correa seemed cognizant of this when he announced the proposal in 2007 in front of the United Nations, noting that "Ecuador seeks to transform old notions of economics and the concept of value...The Yasuní-ITT Project is based on the recognition of use and service of non-chrematistic values of environmental security and maintenance of world biodiversity."[93] Oxford scholar Laura Rival adds that "by valuing the Yasuní for its diversity rather than for the market value of its crude oil content, supporters of the proposal have given visibility to a new national patrimony and opened new possibilities for redefining the forms of social consciousness, collective identity, and will to act through which both nationalism and internationalism get realized in practice." [94] Correa's and Rival's remarks suggest that forests, nature reserves, and indigenous communities can be priceless; they can possess value extending far beyond what we can drill, dam, or dig out of them.

10
Conclusion – Conceptualizing Energy Justice

Introduction

During the 1850s in the United States, when the debate over whether Christianity justified slavery reached its peak, then Illinois Congressperson Abraham Lincoln was reputed to place a silver dollar on top of the Bible and to say during debates that "no man can see the word of justice when it's covered by a silver dollar." His point was that as long as people have a vast economic stake in some sort of existing infrastructure or system, no matter how immoral it may be, they will tend to support it.

Such a stark conclusion may explain why we as a global civilization continue to make little progress addressing our collective challenges relating to energy security and climate change. We are each in our own way complicit with the injustices present in our global energy system, for many of us adhere to values that are inherently materialistic (valuing commodities above people and nature), anthropocentric (concerned primarily with humans), and contempocentric (preoccupied with the short term instead of the long term). As a consequence, most of us fail to properly appreciate how our energy decisions erode the intrinsic worth or vitality of other human beings, ecosystems, and future generations.[1]

This chapter presents a framework, albeit a preliminary one, that tries to alter the values underpinning our energy decisions. It argues that we need to start making energy decisions that promote availability, affordability, due process, good governance, prudence, intergenerational equity, intragenerational equity, and responsibility. Table 10.1 presents these eight core principles as well as the energy-related challenges they address, technology and policy solutions that promote them, and particular case studies investigated in this book's earlier chapters. While all eight principles are important, the idea was to start with the simplest and most accepted ones, such as availability and affordability, before moving towards the more controversial or complex ones such as intragenerational equity and responsibility.

Table 10.1 Energy justice conceptual framework

Principle	Explanation	Challenge	Solution	Case Study
Availability	People deserve sufficient energy resources of high quality	Dependence on foreign suppliers and poor reliability	Distributed generation (including renewable resources and combined heat and power), energy efficiency, carbon taxes	Danish Energy Policy
Affordability	All people, including the poor, should pay no more than ten percent of their income for energy services	Fuel poverty	Home energy efficiency schemes	England's Warm Front Program
Due process	Countries should respect due process and human rights in their production and use of energy	Human rights abuses and flawed impact assessments	Inspection panels	World Bank's Inspection Panel
Information	All people should have access to high-quality information about energy and the environment and fair, transparent, and accountable forms of energy decision-making	Corruption and the resource curse	Transparency and accountability mechanisms	The Extractive Industries Transparency Initiative
Prudence	Energy resources should not be depleted too quickly	Resource exhaustion	Natural Resource Funds	São Tomé e Príncipe's Oil Revenue Management Law
Intergenerational equity	Future generations have a right to enjoy a good life undisturbed by the damage our energy systems inflict on the world today	Energy poverty	Small-scale off-grid renewable sources of energy	IDCOL's Solar Home System program in Bangladesh
Intragenerational equity	People have a right to access energy services fairly	Climate change	Community-based adaptation	GEF's Least Developed Countries Fund
Responsibility	All nations have a responsibility to protect the natural environment and minimize energy-related environmental threats	Externalities	Ecosystem payment schemes and global carbon credits	Ecuador's Yasuni-ITT initiative

After presenting these principles, this chapter concludes by noting some potentially fruitful areas of further research and by calling on energy producers and consumers to recognize the elemental moral and ethical implications of their actions.

Conceptualizing an energy justice framework

Drawn largely from the preceding eight chapters, this section proposes that energy justice consists of the following eight distinct principles:

Availability

Availability is the most basic element of an energy justice framework, for it involves the ability of an economy, market, or system to provide sufficient energy resources when needed. It therefore transcends concerns related to security of supply, sufficiency, and reliability, and it encompasses a range of different dimensions. It includes the physical resource endowment of a particular country or region, as well as the technological solutions that region utilizes to produce, transport, conserve, store, or distribute energy. It includes the amount of investment needed to keep the system functioning, essentially having a robust and diversified energy value chain, as well as promoting infrastructure that can withstand accidental and intentional disruption.[2] Availability is also comparative in the sense that it relates to the relative independence of a country's ability to procure energy fuels and services. Threats to availability can take a variety of forms, from underinvestment in energy infrastructure and burgeoning dependence on imported energy supplies to reliability problems stemming from natural disasters or terrorist attacks. Strategic investments in energy efficiency, renewable energy, and distributed generation can all enhance the availability of energy, and do so in ways that do not conflict with the other justice principles elaborated below.

Affordability

A second core element of energy justice is the basic affordability of energy services, a term that means not just lower prices so that people can afford warm homes and well-lit dwelling spaces, but also energy bills that do not overly burden consumers. The enormous price spikes for natural gas, coal, oil, and uranium seen over the past few years have made many energy facilities costly to operate, and have resulted in significant increases in prices in several areas. Affordability thus encompasses stable prices (minimal volatility) as well as equitable prices that do not require lower-income households to expend disproportionally larger shares of their income on essential services. Schemes that promote better-insulated and energy-efficient homes are one way to enhance affordability.

Due process

The notion of due process – respect for human rights and basic elements of national and international law – dates back thousands of years, yet abuses of those rights and impact assessments that do not adequately protect people are rampant within the energy sector. From an energy justice perspective, due process seeks to ensure stakeholder participation in the energy policymaking process. It also necessitates effective recourse through judicial and administrative remedies and forms of redress. More specifically, the principle suggests that communities must be involved in deciding about projects that will affect them; they must be given fair and informed consent; environmental and social impact assessments must involve genuine community consultation; and neutral arbitration should be available to handle grievances. Inspection panels are one way to promote due process.

Information

This principle suggests that, to minimize corruption and improve account-ability, all people should have access to high-quality information about energy and the environment. Information has become a central element of promoting "good governance" throughout a variety of sectors, a term that centers on democratic and transparent decision-making processes and financial accounting, as well as effective measures to reduce corruption and publish information about energy revenues and policies.[3] Access to infor-mation and transparent frameworks for preserving that access have been known under certain conditions also to encourage democracy, increase busi-ness confidence, and enhance social stability. Voluntary measures such as EITI are one way of enhancing information.

Prudence

Prudence refers to the duty of states to ensure the sustainable use of natural resources. It means that countries have sovereign rights over their natural resources, that they have a duty not to deplete them too rapidly, and that they do not cause undue damage to their environment or that of other states beyond their jurisdiction. Ecologist Paul Hawken eloquently summed up prudence when he wrote that it involves achieving a state where "the demands placed upon the environment by people and commerce can be met without reducing the capacity of the environment to provide for future generations. It can also be expressed in the simple terms of an economic golden rule for the restorative economy: Leave the world better than you found it, take no more than you need, try not to harm life or the environ-ment, make amends if you do."[4] Natural resource funds are one tool that can enhance prudence.

Intergenerational equity

Intergenerational equity – that people have a right to access energy services fairly – finds its roots in modern theories of distributive justice. Philosophers call it "distributive" justice because it deals intently with three aspects of distribution:

1. What goods, such as wealth, power, respect, food, or clothing, are to be distributed?
2. Between what entities are they to be distributed?
3. What is the proper mode of distribution – based on need, based on merit, based on property rights, or something else?

Distributive justice argues that, if physical security is a basic right, then so are the conditions that create it, such as employment, food, shelter, and also unpolluted air, water, and other environmental goods. People are, therefore, entitled to a certain set of minimal energy services which enable them to enjoy a basic minimum of wellbeing. The principle of intergenerational equity aligns with ethicist Henry Shue's wry observation that "whatever justice may positively require, it does not permit the poor nations to be told to sell *their* blankets in order that rich nations may keep *their* jewelry."[5] Small-scale renewable energy technologies such as solar home systems are one tool to enhance the accessibility of energy and thus meet our intergenerational equity obligations.

Intragenerational equity

Instead of emphasizing distributive justice between different communities, intragenerational equity is about distributive justice between present and future generations. It holds that future people have a right to enjoy a good life just like us contemporaries, yet one undisturbed by the temporal damage our energy systems will inflict over time. Consequently, each of us has a moral responsibility to ensure that today's children and future generations inherit a global environment at least no worse than the one we received from our predecessors – and that responsibility extends to preventing climate change and making strategic investments in something known as "adaptation" to increase the resilience of communities. Or, as one commentator put it, "Whatever my attitude toward foreigners, it would be wrong of me to send one a letter-bomb. Would it also not be wrong to send posterity a time-bomb?"[6]

Responsibility

The final principle of responsibility holds that nations have a responsibility to protect the natural environment and minimize the production of negative externalities, or energy-related social and environmental costs. This element of energy justice is perhaps the most controversial and complex, as it blends together four somewhat different notions of "responsibility": a responsibility of governments to minimize environmental degradation, a responsibility of industrialized countries responsible for climate change to pay to fix the problem (the so-called "polluter pays principle"), a responsibility of current generations to protect future ones, and a responsibility of humans to recognize the intrinsic value of non-human species, adhering to a sort of "environmental ethic." Ecosystem payment schemes are one way of adhering to the principle of responsibility.

Further research

Although the eight principles introduced above resonate strongly with me, and I believe them to be salient regardless of whether they are popular, some preliminary work on energy perceptions and attitudes indicates that others find them important as well. With a team of researchers, I recently surveyed 2,167 energy users chosen to represent a variety of government, civil society, private sector, and university stakeholders across the ten countries of Brazil, China, Germany, India, Kazakhstan, Japan, Papua New Guinea, Saudi Arabia, Singapore, and the United States.[7] These respondents were all asked to rate different aspects of energy security and sustainability according to a five-point scale, with "1" indicating "extremely unimportant" and "5" indicating "extremely important."

Despite the diversity of their background, education, age, gender, employment, and culture, Table 10.2 and Figure 10.1, which present the survey's results, show that participants rated *all* eight energy justice principles as "somewhat important" or "extremely important." The amount to which they rated them "important" also differed only slightly, with the lowest (due process) receiving an average score of 4.32 and the highest (responsibility) receiving an average score of 4.58. The implication is that each energy justice principle iterated in my framework is not only "right"; they matter collectively – that is, they matter as a complete framework where every principle has value – and they resound powerfully with a wide range of energy consumers and producers.

However, the work portrayed in this book, and the survey conducted above, can and should be supplemented with at least four types of future inquiry that would help advance our understanding of energy justice.

First and foremost, given the hundreds and perhaps thousands of cases of energy injustice (and efforts to resolve those injustices), the book's sample size of eight cases is remarkably small. The ten countries sampled for the energy security and justice survey also pale in comparison to the almost 200 energy-using countries in the world. Expanding the analysis beyond these case studies and countries to confirm or disprove my eight energy justice principles would be expedient.

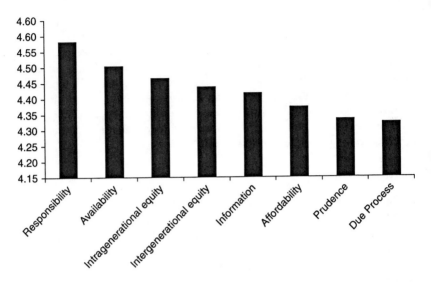

Figure 10.1   Overall mean international ranking of different energy justice principles (n=2,167)

Source: Adapted from Sovacool *et al.*

Second, as stated previously, each of the energy justice principles advanced in this book is weighted equally. That is, availability and affordability carry the same import as equity and responsibility. Yet certain principles may be more salient than others, or they may be prerequisites – one cannot think about energy information and energy due process in a country that has no available energy resources. Consequently, meeting basic energy and subsistence needs is a likely prerequisite for achieving most of the other justice principles. Indeed, research in ethics and social justice,[8] psychology,[9] and household energy security[10] has strongly suggested that subsistence needs must be fulfilled before others can even be considered. Moreover, some energy injustices are certainly more severe than others. A person suffering from fuel poverty or lack of access to modern forms of energy can literally die from freezing to death in the winter or see their life dramatically shortened from indoor air pollution, whereas one suffering from lack of due process can more often than not live to fight another day. The implication is that energy justice principles carry different weights, though how those weights ought to be applied is a complicated topic worthy of future research.

Third, and similarly, research distinguishing justice inputs – things going "into" energy decision-making – from outputs – what society gets "out" of just energy decisions – would be valuable. It could be that some criteria, such as information, equitable participation, and due process, are fruitful inputs, ingredients that result in energy-just decisions, whereas affordability

Table 10.2 International ranking of different energy justice principles by country (n=2,167)

Energy Justice Dimension	Brazil	China	Germany	India	Japan	Singapore	United States	Papua New Guinea	Saudi Arabia	Kazakhstan	Total
Responsibility	4.59	4.70	4.90	4.59	4.56	4.63	4.69	4.49	4.48	4.53	4.58
Availability	4.42	4.89	3.88	4.69	4.62	4.63	4.17	4.49	4.65	4.71	4.50
Intragenerational equity	4.55	4.62	4.36	4.43	4.42	4.62	4.59	4.51	4.41	4.32	4.47
Intergenerational equity	4.50	4.43	4.39	4.33	4.30	4.62	4.57	4.61	4.58	4.41	4.44
Information	4.53	4.10	4.57	4.57	4.29	4.52	4.59	4.59	4.58	4.39	4.42
Affordability	4.53	4.27	4.30	4.51	4.54	4.57	4.13	4.61	4.48	4.53	4.37
Prudence	4.40	4.64	4.21	4.45	4.57	3.78	4.11	4.23	4.27	4.56	4.33
Due Process	4.37	4.28	4.30	4.42	4.19	4.28	4.51	4.59	4.33	4.39	4.32

Source: Adapted from Sovacool *et al.*

and equity are fruitful outputs, things that energy justice bestows on society. My framework above functionally mixes inputs and outputs together, but demarcating them would be helpful.

Fourth, and perhaps most challenging, is the issue of causality. I relied on a rudimentary notion of causality that presumed a particular policy, program, or project accomplished its results if things improved over the duration of that case study. In other words, the Warm Front Scheme in England is presumed to have been responsible for household improvements in energy efficiency even if other local and global forces were at work, including historic lows for energy prices in England. I similarly presumed that the IDCOL program in Bangladesh was the cause of households adopting solar home systems even though at least a dozen other national and bilateral programs were at work disseminating the same technology. Research determining causality in such complex policy environments and energy marketplaces would be exceedingly beneficial.

Closing thoughts

In looking over the previous chapters, I suppose it is fitting to save the Yasuní-ITT initiative for last because it is innovative and in many ways subversive (it alters the fundamental value of oil from something to be combusted to something best left alone underground), and because it cuts to the core of what energy justice is about: the hidden moral and ethical values behind the decisions we make about energy production and use. We can choose to adopt the view that markets adequately distribute energy fuels and services; continue to value lumps of coal, barrels of oil, and cubic meters of natural gas for the motive power or electricity they give us through dollars and cents; and place the responsibility for delivering them in the hands of governments and energy companies.

Or we can adopt a broader view that markets are inherently amoral; that sees a global energy system replete with extreme injustices and asymmetries; and that acknowledges we are each energy agents as much as we are energy consumers. Under this view, we are all guilty in perpetuating energy injustices through our own demand for, and consumption of, energy with tremendously negative, though often surreptitious, impacts on other people and the natural environment. Lumps of coal become acid rain bleaching a stream or forest; barrels of oil become particulates being inhaled by children living near a refinery; and cubic meters of natural gas become responsible for tidal inundations and storm surges flooding our low-lying cities, all producing damage that cannot be entirely measured in monetary value.

The first step towards breaking out of our present global energy system is perhaps recognizing the fundamental injustice of it all, taking a degree of responsibility for it, and embarking on a course of correction and rehabilitation. We can pursue only those energy options that meet the principles of availability, affordability, due process, information, prudence, equity, and responsibility sprinkled throughout this book. For choosing to ignore the ethical implications of our energy system is not a decision free from value; in the struggle between those oppressed by patterns of energy production and use, and those benefitting from that exploitation, doing nothing sides with and validates the oppressive system. It is not to be neutral.

Notes

1 Introduction

1. Leopold, A. *A Sand County Almanac* (Oxford, UK: Oxford University Press; 1949).
2. Abramsky, K. "Introduction: Racing to 'Save' the Economy and the Planet," in K. Abramsky (Ed.) *Sparking a Worldwide Energy Revolution: Social Struggles in the Transition to a Post-Petrol World* (Oakland: AK Press; 2010, pp. 5–30).
3. Goodman, P. "Can Technology Be Humane?" in A. H. Teich (Ed.) *Technology and the Future* (8th ed.) (Boston: Bedford/St Martin's; 2000, pp. 90–102).
4. Noble, D. F. *The Religion of Technology: The Divinity of Man and the Spirit of Invention* (New York: Penguin Books; 1997, p. 208).
5. Verrastro, F. and S. Ladislaw. "Providing Energy Security in an Interdependent World," *The Washington Quarterly* 2007; 30(4): 95–104.
6. Davis, S. J., G. P. Peters and K. Caldeira, "The Supply Chain of CO_2 Emissions," *PNAS* 2011; 108(45): 18554–9.
7. This is not to say, however, that promoting energy justice always makes money, or that our other case studies have similar positive cost curves. Many of the cases to come, such as the World Bank's Inspection Panel, have (deservedly) stopped millions of dollars of energy projects from going forward. The improved transparency from the EITI is undoubtedly losing some oil contractors money. And Denmark had to funnel billions of tax dollars into energy research and development before it became the world leader in wind turbine technology.
8. Vidal, J. "No New Coal—The Calling Card of the 'Green Banksy' who Breached Fortress Kingsnorth,' *The Guardian*, December 11, 2008. Available at http://www.guardian.co.uk/environment/2008/dec/11/kingsnorth-green-banksy-saboteur (accessed: February 1, 2013)
9. Barbeler, D. "Greenpeace ramps up coal protests," *The Courier Mail* (Australia), August 6, 2009.
10. Sovacool, B. K., S. Dhakal, O. Gippner and M. J. Bambawale. "Halting Hydro: A Review of the Socio-Technical Barriers to Hydroelectric Power Plants in Nepal," *Energy* 2011; 36(5): 3468–76.
11. Abramsky, K. "Some Brief News Reports from Direct Action-Based Resistance," in K. Abramsky (Ed.) *Sparking a Worldwide Energy Revolution: Social Struggles in the Transition to a Post-Petrol World* (Oakland: AK Press; 2010, pp. 482–5).
12. Abramsky, p. 483.
13. Abramsky, p. 485.
14. Abramsky, p. 524.
15. Piller, C. *The Fail-Safe Society: Community Defiance and the End of American Technological Optimism* (Los Angeles: University of California Press; 1991, pp. 258–9).
16. Smil, V. "Energy in the Twentieth Century: Resources, Conversions, Costs, Uses, and Consequences," *Annual Review of Energy and Environment* 2000; 25: 21–51.
17. Hall, C., P. Tharakan, J. Hallock, C. Cleveland and M. Jefferson, "Hydrocarbons and the Evolution of Human Culture," *Nature* 2003; 426: 318–22.
18. Sovacool, B. K. *The Dirty Energy Dilemma* (Westport: Praeger; 2008).

19. Kohler, B. "Sustainability and Just Transition in the Energy Industries," in K. Abramsky (Ed.) *Sparking a Worldwide Energy Revolution: Social Struggles in the Transition to a Post-Petrol World* (Oakland: AK Press; 2010, pp. 569–76).

20. Goldthau, A. and B. K. Sovacool. "The Uniqueness of the Energy Security, Justice, and Governance Problem," *Energy Policy* 2012; 41: 232–40.

21. International Energy Agency. *World Energy Outlook 2011* (Paris: OECD; 2011, p. 2). See also International Energy Agency. *Key World Energy Statistics 2011* (Paris: OECD; October).

22. Ibid.

23. Energy Information Administration (EIA). *An Updated Annual Energy Outlook 2009 Reference Case Reflecting Provisions of the American Recovery and Reinvestment Act and Recent Changes in the Economic Outlook,* SR/OIAF/2009–03 (Washington, DC: DOE; 2009, Tables A1 and A8). Available at http://www.eia.doe.gov/oiaf/servicerpt/stimulus/excel/aeostimtab_18.xls (accessed: February 1, 2013).

24. Sovacool, B. K. *The Dirty Energy Dilemma: What's Blocking Clean Power in the U.S.* (Westport, CT: Praeger; 2008, p. 17).

25. Myers, N. and J. Kent. *Perverse Subsidies: How Tax Dollars can Undercut the Environment and the Economy* (Washington: Island Press; 2001).

26. Shields, W. M. "The automobile as an open to closed technological system. Theory and practice in the study of technological systems" (PhD dissertation, Virginia Polytechnic Institute and State University, 2007).

27. Goldthau and Sovacool (2012).

28. International Energy Agency. *Key World Energy Statistics 2011* (Paris: OECD; October).

29. Federal Highway Administration, http://www.fhwa.dot.gov/ohim/tvtw/tvtpage.cfm (accessed: February 1, 2013).

30. The National Railroad Passenger Corporation, known as Amtrak, began operation in 1971. Amtrak revenue-passenger miles have grown at an average annual rate of 2.9 percent from 1971 to 2006, rising to 5.4 billion passenger-miles in 2005. Commuter rail grew to about 9.5 billion passenger-miles in 2005, and rail transit passenger-miles grew to 16 billion in 2005 (Davis, S. C., S. W. Diegel and R. G. Boundy. *Transportation Data Book: Edition 27,* ORNL-6081 (Oak Ridge: Oak Ridge National Laboratory; 2008, Tables 9.10–9.12)). In total, these three rail transportation modes represented 30.9 billion passenger-miles in 2005. At the same time, vehicle-miles per capita grew to 10,082 in 2005 (Ibid., *Transportation Data Book: Edition 27,* Table 8.2). This amounts to 3.2 trillion miles, based on a US population of 296 million in 2005. Thus, the US total passenger-miles on Amtrak, commuter rail, and rail transit represent less than 1 percent of the total vehicle-miles traveled by US passengers in 2005.

31. EIA. *Annual Energy Outlook 2009,* Tables A2 and A17.

32. Sperling, D. and D. Gordon, *Two Billion Cars: Driving Toward Sustainability* (New York, NY: Oxford University Press; 2009).

33. Sperling and Gordon. *Two Billion Cars: Driving Toward Sustainability.*

34. Olz, S., R. Sims and N. Kirchner. Contributions of Renewables to Energy Security. *International Energy Agency Information Paper.* Paris: OECD; 2007.

35. World Bank 2011. *Household Cookstoves, Environment, Health & Climate Change: A New Look at an Old Problem* (Washington, DC: World Bank; 2011).

36. Crewe, E., S. Sundar and P. Young. *Building a Better Stove: The Sri Lanka Experience.* Colombo, Sri Lanka: Practical Action; 2010.

37. World Health Organization 2005.

38. The World Bank. 2012 *Data.* Available at: http://data.worldbank.org/indi-cator/EN.ATM.CO2E.KT?order=wbapi_data_value_2008+wbapi_data_value+wbapi_data_value-last&sort=asc (accessed: January 23, 2012).

39. See Biswas, W. K., P. Bryce and M. Diesendorf. "Model for empowering rural poor through renewable energy technologies in Bangladesh," *Environmental Science & Policy* 2001; 4: 333–44; Krishnan, R. 2009. "Towards Energy Security: Challenges and Opportunities for India," *Paper Presented to the Emerging Challenges to Energy Security in the Asia Pacific International Seminar* (Chennai, India: Center for Security Analysis, March 16 and 17, 2009); Islam, K. R. and R. R. Weil. "Land use effects on soil quality in a tropical forest ecosystem of Bangladesh," *Agriculture, Ecosystems and Environment* 2000; 79: 9–16; Miah, D., H. Al Rashid and M. Y. Shin. "Wood Fuel Use in the Traditional Cooking Stoves in the Rural Floodplain Areas of Bangladesh: A Socio-Environmental Perspective," *Biomass and Bioenergy* 2009; 33: 70–8. Kammen, D. M. and M. R. Dove. "The Virtues of Mundane Science," *Environment* 1997; 39(6): 10–41.

40. Brown, M. A. and B. K. Sovacool. "Developing an 'Energy Sustainability Index' to Evaluate Energy Policy," *Interdisciplinary Science Reviews* 2007; 32(4): 335–49.

41. See Sovacool, B. K. and M. A. Brown. "Competing Dimensions of Energy Security: An International Review," *Annual Review of Environment and Resources* 2010; 35: 77–108; Sovacool, B. K. and M. A. Brown. "Measuring Energy Security Performance in the OECD," in B. K. Sovacool (Ed.) *The Routledge Handbook of Energy Security* (London: Routledge; 2010, pp. 381–95).

42. Sovacool, B. K., I. Mukherjee, I. M. Drupady and A. L. D'Agostino. "Evaluating Energy Security Performance from 1990 to 2010 for Eighteen Countries," *Energy* 2011; 36(10): 5846–53.

43. Sovacool *et al.* (2011).

44. Goldwyn, D. L. (Ed.) *Drilling Down: The Civil Society Guide to Extractive Industry Revenues and the EITI* (Washington, DC: Revenue Watch; 2008).

45. United Nations Development Program, *Yasuní ITT FAQs* (2011).

2 Availability and Danish Energy Policy

1. Portions of this case study are based on Sovacool, B. K, H. H. Lindboe and O. Odgaard. "Is the Danish Wind Energy Model Replicable for Other Countries?" *Electricity Journal* 2008; 21(2): 27–38; and Brown, M. A. and B. K. Sovacool, *Climate Change and Global Energy Security: Technology and Policy Options* (Cambridge: MIT Press; 2011).

2. Sovacool, B. K. and M. A. Brown. "Competing Dimensions of Energy Security: An International Review," *Annual Review of Environment and Resources* 2010; 35: 77–108.

3. Grubb, M., L. Butler and P. Twomey. "Diversity and Security in UK Electricity Generation: The influence of low-carbon objectives," *Energy Policy* 2006; 34: 4050–62; Stirling, A. "Multicriteria Diversity Analysis: A Novel Heuristic Framework for Appraising Energy Portfolios," *Energy Policy* 2010; 38(4): 1622–34; Vivoda, V. "Diversification of Oil Import Sources and Energy Security: A Key Strategy or Elusive Objective?" *Energy Policy* 2009; 37(11): 4615–23.

4. Rogner, H.-H., L. M. Langlois, A. McDonald, D. Weisser and M. Howells. *The Costs of Energy Supply Security* (Vienna: International Atomic Energy Agency; 2006).

5. Leiby, P. N., D. W. Jones, T. R. Curlee and R. Lee. *Oil Imports: An Assessment of Benefits and Costs* (Oak Ridge: Oak Ridge National Laboratory; 1997, ORNL-6851);

Jones, D. W., P. N. Leiby and I. K. Paik. "Oil Price Shocks and the Macroeconomy: What Has Been Learned Since 1996?" *The Energy Journal* 2004; 25(2): 1–32.
6. Loschel, A., U. Moslener and D. Rubbelke. "Indicators of Energy Security in Industrialized Countries," *Energy Policy* 2010; 38: 1665–71.
7. Wilen, J. "Oil Rises Slightly on Pipeline Fire," Associated Press, November 29, 2007.
8. CNA. *Powering America's Defense: Energy and the Risks to National Security* (Alexandria: CNA Corporation; 2009).
9. Greene, D. and S. Ahmad. *Costs of U.S. Oil Dependence: 2005 Update* (January, 2005). Report to the US DOE, ORNL/TM-2005/45.
10. Greene, D. L. "Measuring Energy Security: Can the United States Achieve Oil Independence," *Energy Policy* 2010; 38: 1614–21.
11. Smil, V. *Energy Myths and Realities: Bringing Science to the Energy Policy Debate* (Washington, DC: Rowman and Littlefield; 2010, p. 3).
12. O'Hanlon, M. "How Much does the United States Spend Protecting Persian Gulf Oil?" in C. Pascual and J. Elkind (Eds) *Energy Security: Economics, Politics, Strategies and Implications* (Washington, DC: Brookings Institution Press; 2010, pp. 59–72.
13. Delucchi, M. A. and J. J. Murphy. "US military expenditures to protect the use of Persian Gulf oil for motor vehicles," *Energy Policy* 2008; 36: 2253–64.
14. Luft, G. "United States: a Shackled Superpower." In G. Luft and A. Korin (Eds) *Energy Security Challenges for the 21st Century* (Denver, Colorado: Praegar International/ABC-CLIO; 2009, pp. 143–59).
15. IEA, *World Energy Outlook 2008.*
16. Sovacool, B. K. "Sound Climate, Energy, and Transport Policy for a Carbon Constrained World," *Policy & Society* 2009; 27(4): 273–83.
17. Bacon, R. and M. Kojima. *Vulnerability to Oil Price Increases: A Decomposition Analysis of 161 Countries.* Washington, DC: World Bank, Extractive Industries for Development Series; 2008.
18. Sovacool, B. K. *The Dirty Energy Dilemma: What's Blocking Clean Power in the UnitedStates* (Westport, CT: Praegar; 2008).
19. Lovins and Lovins. *Brittle Power.*
20. United Nations Development Programme (2010).
21. The Chairman of the Presidential Committee on Power Sector Reforms, Dr Rilwanu Lukman, indicates that "the average age of all the transformers, generating stations and sub-stations in the country is twenty five years;" the aging infrastructure has been made worse again, he said, by a "poor maintenance culture." Dr Lukman estimates that $85 billion dollars is needed to overhaul and fix the decrepit power sector, a sum that does not even include related gas infrastructure needs. "$85 Billion Needed for Stable Power; Supply Won't Improve Till December;" Daily Trust/All Africa Global Media, June 25, 2008.
22. World Bank. *Project Paper Proposed Additional Financing Credit in the Amount of SDR $49.6 Million and a Proposed Additional Financing Grant in the Amount of SDR $10.5 Million to Nepal for the Power Development Project* (Washington, DC: World Bank Group; May 18, 2009, Report No. 48516-NP).
23. Mortensen, H. C. and B. Overgaard. "CHP development in Denmark: Role and results," *Energy Policy* 1992; December: 1198–1206.
24. Elliott, R. N. and M. Spurr. *Combined Heat and Power: Capturing Wasted Energy* (Washington, DC: ACEEE; May, 1999).
25. Sovacool and Watts (2009).
26. Lund, H. and B. V. Mathiesen. "Energy System Analysis of 100% Renewable Energy Systems: The Case of Denmark in Years 2030 and 2050," *Energy* 2009; 34: 524–31.

27. Such as waste incineration and the use of biomass.
28. Möller, B. "Spatial analyses of emerging and fading wind energy landscapes in Denmark," *Land Use Policy* 2010; 27: 233–41.
29. Maegaard, P. "Denmark: Politically Induced Paralysis in Wind Power's Homeland and Industrial Hub," in K. Abramsky (Ed.) *Sparking a Worldwide Energy Revolution: Social Struggles in the Transition to a Post-Petrol World* (Oakland: AK Press; 2010, pp. 489–94).
30. Mendonca, M., S. Lacey and F. Hvelplund. "Stability, participation and transparency in renewable energy policy: Lessons from Denmark and the United States," *Policy and Society* 2009; 27: 379–98.
31. Morthorst, P. E. "The Development of a Green Certificate Market," *Energy Policy* 2000; 28: 1085–94.
32. Maegaard, P. "Transition to an Energy-Efficient Supply of Heat and Power in Denmark," in K. Abramsky (Ed.) *Sparking a Worldwide Energy Revolution: Social Struggles in the Transition to a Post-Petrol World* (Oakland: AK Press; 2010, pp. 292–300).
33. Lund, H. "The implementation of renewable energy systems: Lessons learned from the Danish case," *Energy* 2010; 35: 4003–9; see also Lund, H. "Choice awareness: the development of technological and institutional choice in the public debate of Danish energy planning," *Journal of Environmental Policy & Planning* 2000; 2: 249–59.
34. Mortensen, H. C. and B. Overgaard. "CHP development in Denmark: Role and results," *Energy Policy* 1992; December: 1198–1206.
35. Lund, H. and F. Hvelplund. "Does Environmental Impact Assessment Really Support Technological Change? Analyzing Alternatives to Coal-Fired Power Stations in Denmark," *Environmental Impact Assessment Review* 1997; 17: 357–70.
36. Mortensen and Overgaard (1992).
37. Mortensen and Overgaard (1992).
38. Voytenko, Y. and P. Peck. "Organisational frameworks for straw-based energy systemsin Sweden and Denmark," *Biomass and Bioenergy* 2012; 38: 34–8.
39. Hendriks, C. and K. Blok. "Regulation for Combined Heat and Power in the European Union," *Energy Conversion and Management* 1996; 37(6–8): 729–34.
40. van der Vleuten, E. and R. Raven. "Lock-in and change: Distributed generation in Denmark in a long-term perspective," *Energy Policy* 2006; 34: 3739–48.
41. Lehtonen, M. and S. Nye. "History of electricity network control and distributed generation in the UK and Western Denmark," *Energy Policy* 2009; 37: 2338–45.
42. Münster, M., P. E. Morthorst, H. V. Larsen, L. Bregnbæk, J. Werling, H. H. Lindboe and H. Ravn. "The role of district heating in the future Danish energy system," *Energy* (in press, 2012), pp. 1–9.
43. Lund, H., B. Moller, B.V. Mathiesen and A. Dyrelund. "The role of district heating in future renewable energy systems," *Energy* 2010; 35: 1381–90.
44. Odgaard, O. *Energy Policy in Denmark* (Copenhagen: Danish Energy Authority; 2007, p. 9).
45. Olesen, G. B. *Danish Initiatives and Plans in the Field of Energy Efficiency and Renewable Energy* (Forum for Energy Development/ OVE-Europe and the Danish Organisation for Renewable Energy, INFORSE-Europe, 2006).
46. Many of the following paragraphs are based on Togeby, M., K. Dyhr-Mikkelsen, A. Larsen, M. J. Hansen and P. Bach. "Danish energy efficiency policy: revisited and future improvements," *European Council for an Energy-Efficient Economy 2009 Summer Study*, pp. 299–310.

47. For excellent explorations of the Danish approach to wind research and development, see Sovacool, B. K. and J. Sawin. "Creating Technological Momentum: Lessons from American and Danish Wind Energy Research," *Whitehead Journal of Diplomacy and International Relations* 2010; 11(2): 43–57; Garud, R. and P. Karnoe. "Bricolage versus Breakthrough: Distributed and Embedded Agency in Technology Entrepreneurship," *Research Policy* 2003; 32: 277–300; Heymann, M. "Signs of Hubris: The Shaping of Wind Technology Styles in Germany, Denmark, and the United States, 1940–1990," *Technology & Culture* 1998; 39(4): 641–70; Jorgensen, U. and P. Karnoe. "The Danish Wind-Turbine Story: Technical Solutions to Political Visions?" in Rip, A., T. J. Misa and J. Schot (Eds) *Managing Technology in Society: The Approach of Constructive Technology Assessment* (London: Pinter Publishers; 1995, pp. 57–82); Toke, D., S. Breukers and M. Wolsink. "Wind Power Deployment Outcomes: How Can We Account for the Differences?" *Renewable and Sustainable Energy Reviews* 2008; 12: 1129–47; and Mendonça, M., S. Lacey and F. Hvelplund. "Stability, Participation and Transparency in Renewable Energy Policy: Lessons from Denmark and the United States," *Policy & Society* 2009; 27(4): 379–98.
48. Lehtonen, M. and S. Nye. "History of electricity network control and distributed generation in the UK and Western Denmark," *Energy Policy* 2009; 37: 2338–45.
49. Togeby, M., K. Dyhr-Mikkelsen, A. Larsen, M. J. Hansen and P. Bach. "Danish energy efficiency policy: revisited and future improvements," *European Council for an Energy-Efficient Economy 2009 Summer Study*, pp. 299–310.
50. For more on how the trust functioned independently of government, see Lund, H. "Implementation of energy-conversion policies: the case of electric heating conversion in Denmark," *Applied Energy* 1999; 64(1–4): 117–27.
51. Maegaard, P. "Transition to an Energy-Efficient Supply of Heat and Power in Denmark," in K. Abramsky (Ed.) *Sparking a Worldwide Energy Revolution: Social Struggles in the Transition to a Post-Petrol World* (Oakland: AK Press; 2010, pp. 292–300).
52. Olesen, G. B. *Danish Initiatives and Plans in the Field of Energy Efficiency and Renewable Energy* (Forum for Energy Development/ OVE-Europe and the Danish Organisation for Renewable Energy, INFORSE-Europe, 2006).
53. Lund, H. and B. V. Mathiesen. "Energy System Analysis of 100% Renewable Energy Systems: The Case of Denmark in Years 2030 and 2050," *Energy* 2009; 34: 524–31.
54. Sperling, K., F. Hvelplund and B. V. Mathiesen. "Centralisation and decentralisation in strategic municipal energy planning in Denmark," *Energy Policy* 2011; 39: 1338–51.
55. Lund and Mathieson (2009).
56. Danish Energy Agency. *Energy Efficiency Policies and Measures in Denmark* (Copenhagen; 2009).
57. Olesen, G. B. *Danish Initiatives and Plans in the Field of Energy Efficiency and Renewable Energy* (Forum for Energy Development/ OVE-Europe and the Danish Organisation for Renewable Energy, INFORSE-Europe, 2006).
58. Hamilton, B. *A Comparison of Energy Efficiency Programmes for Existing Homes in Eleven Countries* (Montpelier: Regulatory Assistance Project; February 19, 2010).
59. Parajuli, R. "Looking into the Danish energy system: Lesson to be learnedby other communities," *Renewable and Sustainable Energy Reviews* 2012; 16: 2191–9.
60. Lewis, J. and R. Wiser. *Fostering a Renewable Energy Technology Industry: An International Comparison of Wind Industry Policy Support Mechanisms*, LBNL-59116 (Berkeley, CA: Lawrence Berkeley National Laboratory; 2005).

61. Ladenburg, J. "Attitudes towards on-land and offshore wind power development in Denmark; choice of development strategy," *Renewable Energy* 2008; 33: 111–18.
62. Sovacool, B. K. and M. A. Brown. "Competing Dimensions of Energy Security: An International Review," *Annual Review of Environment and Resources* 2010; 35: 77–108.
63. Maegaard, P. "Denmark: Politically Induced Paralysis in Wind Power's Homeland and Industrial Hub," in K. Abramsky (Ed.) *Sparking a Worldwide Energy Revolution: Social Struggles in the Transition to a Post-Petrol World* (Oakland: AK Press; 2010, pp. 489–94).
64. Sperling, K., F. Hvelplund and B. V. Mathiesen. "Evaluation of wind power planning in Denmark: Towards an integrated Perspective," *Energy* 2010; 35: 5443–54.
65. Lehtonen, M. and S. Nye. "History of electricity network control and distributed generation in the UK and Western Denmark," *Energy Policy* 2009; 37: 2338–45.
66. Akhmatov, V. and H. Knudsen. "Large penetration of wind and dispersed generation into Danish power grid," *Electric Power Systems Research* 2007; 77: 1228–38.
67. Lehtonen, M. and S. Nye. "History of electricity network control and distributed generation in the UK and Western Denmark," *Energy Policy* 2009; 37: 2338–45.
68. Sperling, K., F. Hvelplund and B. V. Mathiesen. "Centralisation and decentralisation in strategic municipal energy planning in Denmark," *Energy Policy* 2011; 39: 1338–51.
69. Maegaard, P. "Denmark: Politically Induced Paralysis in Wind Power's Homeland and Industrial Hub," in K. Abramsky (Ed.) *Sparking a Worldwide Energy Revolution: Social Struggles in the Transition to a Post-Petrol World* (Oakland: AK Press; 2010, pp. 489–94).
70. Pasqualetti, M. J. "Opposing Wind Energy Landscapes: A Search for Common Cause," *Annals of the Association of American Geographers*, 2011; 101(4): 1–11.
71. Möller, B. "Spatial analyses of emerging and fading wind energy landscapes in Denmark," *Land Use Policy* 2010; 27: 233–41.
72. Moller, B. "Changing wind-power landscapes: Regional assessment of visual impact on land use and population in Northern Jutland, Denmark," *Applied Energy* 2006; 83: 477–94.
73. US Energy Information Administration. *Denmark Energy Profile*, available at http://tonto.eia.doe.gov/country/country_energy_data.cfm?fips=DA (accessed: 1 February, 2013).
74. Danish Energy Association. *Danish Electricity Supply Statistical Survey* (Denmark: 2008, pp. 1–3). Available at http://www.danishenergyassociation.com/Statistics.aspx (accessed: 1 February, 2013).
75. Østergaard, P. A. "Regulation strategies of cogeneration of heat and power (CHP) plants and electricity transit in Denmark," *Energy* 2010; 35: 2194–202.
76. Sperling, K., F. Hvelplund and B. V. Mathiesen. "Centralisation and decentralisation in strategic municipal energy planning in Denmark," *Energy Policy* 2011; 39: 1338–51.
77. Tonini, D. and T Astrup. "LCA of biomass-based energy systems: A case study for Denmark," *Applied Energy* 2012; 99: 234–46.
78. Parajuli, R. "Looking into the Danish energy system: Lesson to be learned by other communities," *Renewable and Sustainable Energy Reviews* 2012; 16: 2191–9.
79. Lund, H. "The implementation of renewable energy systems: Lessons learned from the Danish case," *Energy* 2010; 35: 4003–9.

80. van der Vleuten, E. and R. Raven. "Lock-in and change: Distributed generation in Denmark in a long-term perspective," *Energy Policy* 2006; 34: 3739–48.
81. Sperling, K., F. Hvelplund and B. V. Mathiesen. "Centralisation and decentralisation in strategic municipal energy planning in Denmark," *Energy Policy* 2011: 39: 1338–51.
82. Toke, D. "Are Green Electricity Certificates the Way Forward for Renewable Energy? An Evaluation of the United Kingdom's Renewables Obligation in the Context of International Comparisons," *Environment and Planning C* 2005; 23: 361–74.

3   Affordability and Fuel Poverty in England

1. Department for Energy and Climate Change. *Connecting with communities The Warm Front Scheme Annual Report 2010/11*.
2. Sovacool, B. K. and M. A. Brown. "Competing Dimensions of Energy Security: An International Review," *Annual Review of Environment and Resources* 2010; 35: 77–108.
3. Walker, G. and R. Day. "Fuel poverty as injustice: Integrating distribution, recognition and procedure in the struggle for affordable warmth," *Energy Policy* 2012; 49: 69–75.
4. Ibid.
5. Liddell, C., C. Morris, S. J. P. McKenzie and G. Rae. "Measuring and monitoring fuel poverty in the UK: National and regional perspectives," *Energy Policy* 2012; 49: 27–32. See also Liddell, C. "Fuel poverty comes of age: Commemorating 21 years of research and policy," *Energy Policy* 2012; 49: 2–5; Liddell, C. "The missed exam: Conversations with Brenda Boardman," *Energy Policy* 2012; 49: 12–18.
6. Moore, R. "Definitions of fuel poverty: Implications for policy," *Energy Policy* 2012; 49: 19–26.
7. See also Boardman, B. *Fixing Fuel Poverty: Challenges and Solutions* (London: Earthscan; 2010) and Boardman, B. *Fuel Poverty: From Cold Homes to Affordable Warmth* (London: Belhaven; 1991).
8. Boardman, B. "Opportunities and Constraints Posed by Fuel Poverty on Policies to Reduce the Greenhouse Effect in Britain," *Applied Energy* 1993; 44: 185–95.
9. Boardman, B. "Fuel Poverty," *International Encyclopedia of Housing and Home* (London: Elsevier; 2012, pp. 221–5). http://www.sciencedirect.com/science/referenceworks/9780080471716 (accessed: March 20, 2013).
10. Howden-Chapman, P., H. Viggers, R. Chapman, K. O'Sullivan, L. Telfar Barnard and B. Lloyd. "Tackling cold housing and fuel poverty in New Zealand: A review of policies, research, and health impacts," *Energy Policy* 2012; 49: 134–42.
11. Brunner, K.-M., M. Spitzer and A. Christanell, "Experiencing fuel poverty. Coping strategies of low-income households in Vienna/Austria," *Energy Policy* 2012; 49: 53–9.
12. Soviet-style district heating systems tend to heat higher floors in prefabricated block apartments much more than lower floors, to the degree where heat is regulated in the winter by opening windows because it is too hot. Panel apartments become very uncomfortable in the summer and most apartments lack even a control device, instead relying on centralized single-loop heat distribution systems. Hungarian consumers must still pay for this heat – and Hungary reports as much as 50 percent higher annual energy and heating costs per person than the European Union average – and the situation is worsened by the difficulty or

even impossibility of disconnecting from district heating networks or switching to other sources.

13. Tirado Herrero S. and D. Urge-Vorsatz, "Trapped in the heat: A post-communist type of fuel poverty," *Energy Policy* 2012; 49: 60–8.
14. O'Brien, M. "Policy Summary: Fuel Poverty in England," *The Lancet* (December, 2011), available at http://ukpolicymatters.thelancet.com/?p=1603 (accessed: February 1, 2013).
15. Howden-Chapman, P., H. Viggers, R. Chapman, K. O'Sullivan, L. Telfar Barnard and B. Lloyd. "Tackling cold housing and fuel poverty in New Zealand: A review of policies, research, and health impacts," *Energy Policy*, 2012; 49: 134–42.
16. Bradshaw, J. and S. Hutton. "Social Policy Options and Fuel Poverty," *Journal of Economic Psychology* 1983; 3: 249–66.
17. Boardman, B. "Fuel Poverty," *International Encyclopedia of Housing and Home* (2012, pp. 221–5).
18. Tirado Herrero, S. and D. Urge-Vorsatz, "Trapped in the heat: A post-communist type of fuel poverty," *Energy Policy* 2012; 49: 60–8.
19. Liddell, C., C. Morris, S. J. P. McKenzie and G. Rae. "Measuring and monitoring fuel poverty in the UK: National and regional perspectives," *Energy Policy* 2012; 49: 27–32.
20. O'Brien, M. "Policy Summary: Fuel Poverty in England," *The Lancet* (December, 2011), available at http://ukpolicymatters.thelancet.com/?p=1603 (accessed: February 1, 2013).
21. Falagas, M. E., D. E. Karageorgopoulos, L. I. Moraitis, E. K. Vouloumanou, N. Roussos, G. Peppas and P. I. Rafailidis. Seasonality of mortality: the September phenomenon in Mediterranean countries. *Canadian Medical Association Journal* 2009; 181: 484–6.
22. Howden-Chapman, P., H. Viggers, R. Chapman, K. O'Sullivan, L. Telfar Barnard and B. Lloyd. "Tackling cold housing and fuel poverty in New Zealand: A review of policies, research, and health impacts," *Energy Policy*, 2012; 49: 134–42.
23. Data for all countries but the UK from Falagas, M. E., D. E. Karageorgopoulos, L. I. Moraitis, E. K. Vouloumanou, N. Roussos, G. Peppas and P. I. Rafailidis. "Seasonality of mortality: the September phenomenon in Mediterranean countries," *Canadian Medical Association Journal* 2009; 181: 484–6. UK data are from Day, R. and R. Hitchings, " 'Only old ladies would do that': Age stigma and older people's strategies for dealing with winter cold," *Health & Place* 2011; 17: 885–94.
24. For excellent reviews, see Shaw, M. "Housing and Public Health," *Annual Review of Public Health* 2004; 25: 397–418; Wilkinson, P., K. R. Smith, M. Joffe and A. Haines. "Global perspective on energy: health effects and injustices," *Lancet* 2007; 370: 965–78; and Jenkins, D. P. "The value of retrofitting carbon-saving measures into fuel poor social housing," *Energy Policy* 2010; 38: 832–9.
25. Liddell, C. and C. Morris, "Fuel poverty and human health: A review of recent evidence," *Energy Policy* 2010; 38: 2987–97.
26. Marmot Review Team. *The Health Impacts of Cold Homes and Fuel Poverty* (Friends of the Earth, May 2011).
27. Ormandy, D. and V. Ezratty, "Health and thermal comfort: From WHO guidance to housing strategies," *Energy Policy* 2012; 49: 116–21.
28. Ibid.
29. Anderson, W., V. White and A. Finney, "Coping with low incomes and cold homes," *Energy Policy* 2012; 49: 40–52.

30. Critchley, R., J. Gilbertson, M. Grimsley and G. Green, Warm Front Study Group, "Living in cold homes after heating improvements: Evidence from Warm-Front, England's Home Energy Efficiency Scheme," *Applied Energy* 2007; 84: 147–58.

31. These claims do depend on the self-reported data and projections from WF reports and documents being accurate.

32. Gilbertson, J., M. Stevens, B. Stiell and N. Thorogood, "Home is where the hearth is: Grant recipients' views of England's Home Energy Efficiency Scheme (Warm Front)," *Social Science & Medicine* 2006; 63: 946–56.

33. Department of Food and Rural Affairs. *The UK Fuel Poverty Strategy* (London: Defra; 2001).

34. Liddell, C., C. Morris, S. J. P. McKenzie and G. Rae. "Measuring and monitoring fuel poverty in the UK: National and regional perspectives," *Energy Policy*, 2012; 49: 27–32.

35. National Audit Office. *Warm Front: Helping to Combat Fuel Poverty* (London: The Stationery Office; 2003).

36. National Audit Office. *The Warm Front Scheme* (London: The Stationery Office; 2009).

37. Department for Energy and Climate Change. *Connecting with communities The Warm Front Scheme Annual Report 2010/11*.

38. Ormandy, D. and V. Ezratty, "Health and thermal comfort: From WHO guidance to housing strategies," *Energy Policy* 2012; 49: 116–21.

39. Department of Food and Rural Affairs. *The UK Fuel Poverty Strategy* (London: Defra; 2001). See also Guertler, P. "Can the Green Deal be fair too? Exploring new possibilities for alleviating fuel poverty," *Energy Policy* 2012; 49: 91–7.

40. National Audit Office. *The Warm Front Scheme* (London: The Stationery Office; 2009).

41. National Audit Office. *Warm Front: Helping to Combat Fuel Poverty* (London: The Stationery Office; 2003).

42. Department of Environment, Food and Rural Affairs. *Fuel poverty in England: the Government's plan for action* (London: DEFRA; 2004).

43. National Audit Office. *The Warm Front Scheme* (London: The Stationery Office; 2009).

44. Presuming that 127,930 households received assistance in 2011, and that each household saves 11.2 GJ per year for the next 20 years, and WF will save roughly 28,656,320 GJ or 7,960,088,890 kWh. With total costs of £195,000,000 per year, this amounts to savings of about 2.4 pence per kWh, or 3.85 US cents per kWh.

45. Department for Energy and Climate Change. *Connecting with communities The Warm Front Scheme Annual Report 2010/11*.

46. The national health impact evaluation of Warm Front was carried out by The Centre for Regional Economic and Social Research at Sheffield Hallam University in partnership with London School of Hygiene and Tropical Medicine and University College London. It is sometimes collectively referred to as the Warm Front Study Group. See also Critchley, R., J. Gilbertson, M. Grimsley and G. Green, Warm Front Study Group. "Living in cold homes after heating improvements: Evidence from Warm-Front, England's Home Energy Efficiency Scheme," *Applied Energy* 2007; 84: 147–58.

47. Warm Front Evaluation Team, *Warm Front Evaluation* (London School of Hygiene and Tropical Medicine; 2006).

48. Hong, S. H., J. Gilbertson, T. Oreszczyn, G. Green and I. Ridley, the Warm Front Study Group. "Field study of thermal comfort in low-income dwellings in England

before and after energy efficient refurbishment," *Building and Environment* 2009; 44: 1228–36.

49. Oreszczyn, T., S. H. Hong, I. Ridley and P. Wilkinson, the Warm Front Study Group. "Determinants of winter indoor temperatures in low-income households in England," *Energy and Buildings* 2006; 38(3): 245–52.

50. National Audit Office. *The Warm Front Scheme* (London: The Stationery Office; 2009).

51. Anderson, W., V. White and A. Finney. "Coping with low incomes and cold homes," *Energy Policy* 2012; 49: 40–52.

52. Ibid.

53. Ibid.

54. Ormandy, D. and V. Ezratty. "Health and thermal comfort: From WHO guidance to housing strategies," *Energy Policy* 2012; 49: 116–21.

55. Gilbertson, G., M. Grimsley and G. Green, For the Warm Front Study Group. "Psychosocial routes from housing investment to health: Evidence from England's home energy efficiency scheme," *Energy Policy* 2012; 49: 122–33.

56. Gilbertson, J., M. Stevens, B. Stiell and N. Thorogood. "Home is where the hearth is: Grant recipients' views of England's Home Energy Efficiency Scheme (Warm Front)," *Social Science & Medicine* 2006; 63: 946–56.

57. According to the latest data available in the WF's Annual Report, 2.3 million households received assistance from WF over the course of 2001 to 2011. According to them "the average annual increase per customer identified" under a benefit entitlement check amounted to £1,894.79 individually or £4,358,017,000 as a whole (if one multiplies that amount for all 2.3 million households), rising to £87,160,340,000 when multiplied over the duration of 20 years. It must be noted, however, that this calculation is very rough. Benefits reported in the Annual Report are entirely theoretical and computed using a method called the Standard Assessment Protocol. Actual savings will vary by household, and would not take into consideration the rebound effect discussed in the Challenges section.

58. Green, G. and J. Gilbertson. *Warm Front, Better Health: Health Impact Evaluation of the Warm Front Scheme* (Center for Regional, Economic, and Social Research; May, 2008).

59. Shortt, N. and J. Rugkasa. " 'The walls were so damp and cold' fuel poverty and ill health in Northern Ireland: Results from a housing intervention," *Health & Place* 2007; 13: 99–110.

60. Ormandy, D. and V. Ezratty. "Health and thermal comfort: From WHO guidance to housing strategies," *Energy Policy* 2012; 49: 116–21.

61. Department for Energy and Climate Change. *Annual report on fuel poverty statistics* (London: National Statistics; 2011).

62. Guertler, P. "Can the Green Deal be fair too? Exploring new possibilities for alleviating fuel poverty," *Energy Policy* 2012; 49: 91–7.

63. Gilbertson, J., M. Grimsley and G. Green, for the Warm Front Study Group. "Psychosocial routes from housing investment to health: Evidence from England's home energy efficiency scheme," Energy Policy 2012; 49: 122–33.

64. Boardman, B. "Fuel Poverty," *International Encyclopedia of Housing and Home* (2012, pp. 221–5).

65. Bahaj, A. S. and P. A. B. James. "Urban energy generation: The added value of photovoltaics in social housing," *Renewable and Sustainable Energy Reviews* 2007; 11: 2121–36.

66. O'Brien, B. "Policy Summary: Fuel Poverty in England," *The Lancet* (December, 2011), available at http://ukpolicymatters.thelancet.com/?p=1603 (accessed: February 1, 2013).
67. Vukmanovic, O. "Quarter of homes seen in fuel poverty by 2015," Reuters News Service, October 10, 2011.
68. Illsley, B., T. Jackson and B. Lynch. "Addressing Scottish rural fuel poverty through a regional industrial symbiosis strategy for the Scottish forest industries sector," *Geoforum* 2007; 38: 21–32.
69. Department for Energy and Climate Change. *Annual report on fuel poverty statistics* (London: National Statistics; 2011).
70. Vukmanovic, O. "Quarter of homes seen in fuel poverty by 2015," Reuters News Service, October 10, 2011.
71. Davies, M. and T. Oreszczyn. "The unintended consequences of decarbonising the built environment: A UK case study," *Energy and Buildings* 2012; 46: 80–5.
72. Guertler, P. "Can the Green Deal be fair too? Exploring new possibilities for alleviating fuel poverty," *Energy Policy* 2012; 49: 91–7.
73. Roberts, S. "Energy, equity and the future of the fuel poor," *Energy Policy* 2008; 36: 4471–4.
74. O'Brien, M. "Policy Summary: Fuel Poverty in England," *The Lancet* (December, 2011), available at http://ukpolicymatters.thelancet.com/?p=1603 (accessed February 1, 2013).
75. Hills, J. *Getting the measure of fuel poverty* (Final Report of the Fuel Poverty Review, Department of Energy and Climate Change, March, 2012).
76. National Audit Office. *The Warm Front Scheme* (London: The Stationery Office; 2009).
77. The WF Scheme helped 127,930 homes in 2011, meaning it saved about 3,837,900 tons of carbon dioxide over a 20-year period at a cost of £195 million, resulting in an abatement cost of £50.81.
78. Sefton, T. "Targeting fuel poverty in England: is the government getting warm?" *Fiscal Studies* 2002; 23(3): 369–99.
79. Sefton, T. *Aiming high: an evaluation of the potential contribution of Warm Front towards meeting the Government's fuel poverty target in England.* CASE report 28 (London, UK: Centre for Analysis of Social Exclusion, London School of Economics and Political Science; 2004).
80. Moore, R. "Definitions of fuel poverty: Implications for policy," *Energy Policy* 2012; 49: 19–26.
81. Hills, J. *Getting the measure of fuel poverty* (Final Report of the Fuel Poverty Review, Department of Energy and Climate Change, March, 2012).
82. National Audit Office. *Warm Front: Helping to Combat Fuel Poverty* (London: The Stationery Office; 2003).
83. National Audit Office. *The Warm Front Scheme* (London: The Stationery Office, 2009).
84. See Waddams Price, C., K. Brazier and W. Wang. "Objective and subjective measures of fuel poverty," *Energy Policy* 2012; 49: 33–9; as well as Dubois, U. "From targeting to implementation: The role of identification of fuel poor households," *Energy Policy* 2012; 49: 107–15.
85. Ormandy, D. and V. Ezratty, "Health and thermal comfort: From WHO guidance to housing strategies," *Energy Policy* 2012; 49: 116–21.
86. Green, G. and J. Gilbertson. *Warm Front, Better Health: Health Impact Evaluation of the Warm Front Scheme* (Center for Regional, Economic, and Social Research, May, 2008).

87. Hong, S. H., T. Oreszczyn and I. Ridley, for the Warm Front Study Group. "The impact of energy efficient refurbishment on the space heating fuel consumption in English dwellings," *Energy and Buildings* 2006; 38: 1171–81.
88. Green, G. and J. Gilbertson. *Warm Front, Better Health: Health Impact Evaluation of the Warm Front Scheme* (Center for Regional, Economic, and Social Research, May, 2008).
89. Monzani, D. *Green Deal Legislation & Finance: UK energy efficiency financing* (London:Department of Energy and Climate Change; February 2012).
90. International Energy Agency. *The Experience With Energy Efficiency Policies and Programs in IEA Countries: Learning from the Critics* (Paris, France: International Energy Agency; August, 2005).
91. De La Torre, A., P. Fajnzylber and J. Nash. *Low-Carbon Development: Latin American Responses to Climate Change* (Washington, DC: World Bank Group; 2010).
92. Personal correspondence with Prof. David Elliott.
93. Boardman, B. "Fuel Poverty," *International Encyclopedia of Housing and Home* (2012, pp. 221–5).
94. Hills Report (2012).

4 Due Process and the World Bank's Inspection Panel

1. Sovacool, B. *The Dirty Energy Dilemma* (Westport: Praegar; 2008, p. 134).
2. International Bank for Reconstruction and Development and International Development Association, *Accountability at the World Bank: The Inspection Panel at 15 Years* (Washington, DC: World Bank Group; 2009).
3. Ballard, C. and G. Banks, "Resource Wars: The Anthropology of Mining," *Annual Review of Anthropology* 2003; 32: 305.
4. *The Guardian*, "Shell Pays Out $15.5m Over Saro-Wiwa Killing," June 9, 2009.
5. Taylor, I. "China's Oil Diplomacy in Africa," *International Affairs* 2006; 82(5): 937–59.
6. Watts, M. J. "Righteous Oil: Human Rights, The Oil Complex, and Corporate Social Responsibility," *Annual Review of Environment and Resources* 2005; 30: 373–407.
7. Sovacool, *Pacific Affairs* (2009).
8. Sovacool, B. K. and L. C. Bulan. "Behind an Ambitious Megaproject in Asia: The History and Implications of the Bakun Hydroelectric Dam in Borneo," *Energy Policy* 2011; 39(9): 4842–59.
9. Sovacool, B. K. and L. C. Bulan. "Energy Security and Hydropower Development in Malaysia: The Drivers and Challenges facing the Sarawak Corridor of Renewable Energy (SCORE)," *Renewable Energy* 2012; 40(1): 113–29; Sovacool, B. K. and L. C. Bulan. "Meeting Targets, Missing People: The Energy Security Implications of the Sarawak Corridor of Renewable Energy (SCORE) in Malaysia," *Contemporary Southeast Asia* 2011; 33(1): 56–82; Sovacool, B. K. and L. C. Bulan. "They'll Be Dammed: The Sustainability Implications of the Sarawak Corridor of Renewable Energy (SCORE) in Malaysia," *Sustainability Science* (in press, 2012).
10. Ballard, C. *Human Rights and the Mining Sector in Indonesia: A Baseline Study* (International Institute for Environment and Development; October 2001).
11. Myers, S. L. "Lament for a Once-Lovely Waterway," *New York Times*, June 12, 2010; Myers, S. L. "Vital River is Withering, and Iraq Has No Answer," *New York Times*, June 12, 2010.
12. Downing, T. E. *Avoiding New Poverty: Mining-Inducted Displacement and Resettlement* (International Institute for Environment and Development; April, 2002).

13. Hunter, D. "Using the World Bank Inspection Panel to Defend the Interests of Project-Affected People," *Chicago Journal of International Law* 2003; 4: 201–11.
14. Hornberger, J. G. *The Bill of Rights: Due Process of Law* (Future of Freedom Foundation; 2005).
15. Ash, J. "New Nuclear Energy, Risk, and Justice: Regulatory Strategies for an Era ofLimited Trust," *Politics & Policy* 2010; 38(2): 255–84; Adger, W. N., J. Paavola and S. Huq, "Toward Justice in Adaptation to Climate Change," in Adger, W. N., J. Paavola, S. Huq., and M. J. Mace (Eds) *Fairness in Adaptation to Climate Change* (Cambridge: MIT Press; 2006, pp. 1–19); Barry, B. *Justice as Impartiality* (Oxford: Oxford University Press; 1995); Salazar, D. J. and D. K. Alper, "Justice and Environmentalisms in the British Columbia and U.S. Pacific Northwest Environmental Movements," *Society & Natural Resources* 2011; 24(8): 767–84.
16. Weston, B. H., "Climate Change and Intergenerational Justice: Foundational Reflections," *Vermont Journal of Environmental Law* 2008; 9: 375–430; see also Weston, B. H. and T. Bach, *Climate Change and Intergenerational Justice: Present law, Future Law* (Vermont Law School; 2008).
17. Anton, D. K. and D. L. Shelton, *Environmental Protection and Human Rights* (New York: Cambridge University Press; 2011, p. 431).
18. Goodland, R. "Free, Prior and Informed Consent and the World Bank Group," *Sustainable Development Law & Policy* 2004; 4(2): 66–74; UN Permanent Forum on Indigenous Issues (UNPFII). *Report of the International Workshop on Methodologies Regarding Free Prior and Informed Consent and Indigenous Peoples.* Document E/C.19/2005/3, submitted to the Fourth Session of UNPFII, May 16–17, 2005.
19. Goodland (2004).
20. Finer, M., C. N. Jenkins, S. L. Pimm, B. Keane and C. Ross. "Oil and Gas Projects in the Western Amazon: Threats to Wilderness, Biodiversity, and Indigenous Peoples," *PLoS One* 2008; 3(8): 1–9.
21. Commission on Human Rights, Sub-Commission on the Promotion and Protection of Human Rights, Working Group on Indigenous Populations, 22nd session, 19–13 July, 2004, p. 5.
22. Colchester, M. and M. F. Ferrari. *Making FPIC – Free, Prior and Informed Consent – Work: Challenges and Prospects for Indigenous People* (Forest Peoples Project, 4 June, 2007); United Nations. *Free Prior Informed Consent and Beyond: The Experience of IFAD* (Geneva: PFII/2005/WS.2/10; 2005); Salazar, D. J. and D. K. Alper, "Justice and Environmentalisms in the British Columbia and U.S. Pacific Northwest Environmental Movements," *Society & Natural Resources* 2011; 24(8): 767–84.
23. The following paragraphs draw substantially from Dunkerton, K. J. "The World Bank Inspection Panel and its Affect on Lending Accountability to Citizens of Borrowing Nations," *University of Baltimore Journal of Environmental Law* 1995; 5: 226–61; as well as The World Bank. *Annual Report 2012* (Washington, DC: World Bank Group).
24. Clark, D. L. *A Citizen's Guide to the World Bank Inspection Panel* (Washington, DC: Center for International Environmental Law; October, 1999).
25. The World Bank. *Annual Report 2012* (Washington, DC: World Bank Group).
26. Woods, N. "The Challenge of Good Governance for the IMF and the World Bank Themselves," *World Development* 2000; 28(5): 823–41.
27. See http://www.worldbank.org/ibrd (accessed February 1, 2013).
28. World Bank Group. "Energy Data." available from http://go.worldbank.org/ERF9QNT660 (accessed: October, 2012).

29. Notes: 1 IBRD: The International Bank for Reconstruction and Development. 2 IDA: International Development Association. 3 Climate Finance: Includes Carbon Finance, the Clean Technology Fund, the Global Environment Facility and the Program for Scaling Up Renewable Energy in Low Income Countries. 4 Others: These include Guarantees, Special Financing and Recipient-Executed activities. 5 IFC: The International Finance Corporation provides loans, equity, and technical assistance to stimulate private sector investment in developing countries. 6 MIGA: The Multilateral Investment Guarantee Agency (MIGA) provides guarantees against losses caused by noncommercial risks to investors in developing countries.

30. Low Carbon projects include renewable energy projects, energy efficiency projects, and projects that support increased use of cleaner fuels to displace more carbon-intensive ones. Energy Access includes projects aimed at increasing access to electricity services.

31. Pincus, J. R. and J. A. Winters, *Reinventing the World Bank* (Ithaca and New York: Cornell University Press; 2002); World Bank. *World Bank Group Work in Low-Income Countries Under Stress: A Task Force Report* (Washington, DC: World Bank; 2002).

32. World Bank Inspection Panel. *The Inspection Panel: Annual Report July 1 2010 to June 20, 2011* (Washington, DC: International Bank for Reconstruction and Development/World Bank Group; 2011).

33. For more on the history of the formation of the IP, readers should peruse The World Bank. *The World Bank Inspection Panel: The First Four Years (1994–1998)* (Washington, DC: World Bank Group; 1998); Bradlow, D. D. and S. Schlemmer-Schulte. "The World Bank's New Inspection Panel: A Constructive Step in the Transformation of the International Legal Order," *Heidelberg Journal of Law* 1994; 392–415; Bradlow, D. D. "International Organizations and Private Complaints: The Case of the World Bank Inspection Panel," *Virginia Journal of International Law* 1994; 34(3): 555–613; Shihata, I. F. I. *The World Bank Inspection Panel* (New York: Oxford University Press; 1994).

34. International Bank for Reconstruction and Development and International Development Association. *Accountability at the World Bank: The Inspection Panel at 15 Years* (Washington, DC: World Bank Group; 2009).

35. Clark, D. "Understanding the World Bank Inspection Panel," in D. Clark, J. Fox and K. Treakle (Eds) *Demanding Accountability: Civil Society Claims and the World Bank Inspection Panel* (New York: Rowman and Littlefield; 2003, pp. 1–24).

36. The International Bank for Reconstruction and Development and the World Bank. *Accountability at the World Bank: The Inspection Panel 10 Years On* (Washington, DC: World Bank Group; 2003).

37. Fox, J. "Introduction: Framing the Inspection Panel," in D. Clark, J. Fox and K. Treakle (Eds) *Demanding Accountability: Civil Society Claims and the World Bank Inspection Panel* (New York: Rowman and Littlefield; 2003, pp. xi–xxxi).

38. International Bank for Reconstruction and Development and International Development Association (2009).

39. Werlin, H. H. "Helping Poor Countries: A Critique of the World Bank," *Orbis* 2003; Fall: 757–65.

40. Clark, D. "Understanding the World Bank Inspection Panel," in D. Clark, J. Fox and K. Treakle (Eds) *Demanding Accountability: Civil Society Claims and the World Bank Inspection Panel* (New York: Rowman and Littlefield; 2003, pp. 1–24).

41. International Bank for Reconstruction and Development and International Development Association. *Accountability at the World Bank: The Inspection Panel at 15 Years* (Washington, DC: World Bank Group; 2009).

42. Ibid.
43. World Bank Inspection Panel. *The Inspection Panel: Annual Report July 1 2010 to June 20, 2011* (Washington, DC: International Bank for Reconstruction and Development/World Bank Group; 2011).
44. International Bank for Reconstruction and Development and International Development Association (2009).
45. Bradlow, D. D. "International Organizations and Private Complaints: The Case of the WorldBank Inspection Panel," *Virginia Journal of International Law* 1993–1994; 34: 553–614.
46. International Bank for Reconstruction and Development and International Development Association (2009).
47. Utzinger, J., K. Wyss, D. D. Moto, N. D. Yemadji, M. Tanner and B. H. Singer. "Assessing Health Impacts of the Chad-Cameroon Petroleum Development and Pipeline Project: Challenges and a Way Forward," *Environmental Impact Assessment Review* 2005; 25: 63–93.
48. Clark, D., "Understanding the World Bank Inspection Panel," in D. Clark, J. Fox and K. Treakle (Eds) *Demanding Accountability: Civil Society Claims and the World Bank Inspection Panel* (New York: Rowman and Littlefield; 2003, pp. 1–24).
49. International Bank for Reconstruction and Development and International Development Association (2009).
50. Ibid.
51. Ibid.
52. Treakle, K., J. Fox and D. Clark, "Lessons Learned," in D. Clark, J. Fox and K. Treakle (Eds) *Demanding Accountability: Civil Society Claims and the World Bank Inspection Panel* (New York: Rowman and Littlefield; 2003, pp. 247–77).
53. Bradlow, D. D. "International Organizations and Private Complaints: The Case of the WorldBank Inspection Panel," *Virginia Journal of International Law* 1993–1994; 34: 553–614.
54. Clark, D. L. *A Citizen's Guide to the World Bank Inspection Panel* (Washington, DC: Center for International Environmental Law; October, 1999).
55. Treakle, K., J. Fox and D. Clark, "Lessons Learned," in D. Clark, J. Fox and K. Treakle (Eds) *Demanding Accountability: Civil Society Claims and the World Bank Inspection Panel* (New York: Rowman and Littlefield; 2003, pp. 247–77).
56. International Bank for Reconstruction and Development and International Development Association (2009).
57. Udall, L. *World Bank Inspection Panel* (Contributing Paper 126 to the World Commission on Dams, 2000).
58. Fox, J. and K. Treakle, "Concluding Propositions," in D. Clark, J. Fox and K. Treakle (Eds) *Demanding Accountability: Civil Society Claims and the World Bank Inspection Panel* (New York: Rowman and Littlefield; 2003, pp. 279–86).
59. International Bank for Reconstruction and Development and International Development Association (2009).
60. Kim, E. T., "Unlikely Formation: Contesting and Advancing Asian/African 'Indigenousness' at the World Bank Inspection Panel," *New York University Journal of International Law and Policy* 2008–2009; 41: 131–58.
61. Treakle, K., J. Fox and D. Clark, "Lessons Learned," in D. Clark, J. Fox and K. Treakle (Eds) *Demanding Accountability: Civil Society Claims and the World Bank Inspection Panel* (New York: Rowman and Littlefield; 2003, pp. 247–77).
62. Guarascio, F. "International Development Banks Remain Opaque Institutions," October 1, 2012, available at http://www.publicserviceeurope.com/article/2523/

international-development-banks-remain-opaque-institutions#ixzz29V4bbgGy (accessed February 1, 2013).

63. European Investment Bank. *Complaints Mechanism Operating Procedures* (Brussels: EIB; April 2012).

64. Boisson de Chazournes, L. "Public Participation in Decision-Making: The World Bank Inspection Panel," *Studies in Transnational Legal Policy* 1999; 31: 84–94.

65. Herz, S. and A. Perrault. *Bringing Human Rights Claims to the World Bank Inspection Panel* (Washington, DC: Center for International Environmental Law, Bank Information Center, and the International Accountability Project; October, 2009).

66. Treakle, K., J. Fox and D. Clark, "Lessons Learned," in D. Clark, J. Fox and K. Treakle (Eds) *Demanding Accountability: Civil Society Claims and the World Bank Inspection Panel* (New York: Rowman and Littlefield; 2003, pp. 247–77).

67. Ibid.

68. Ibid.

69. Dunkerton, K. J. "The World Bank Inspection Panel and its Affect on Lending Accountability to Citizens of Borrowing Nations," *University of Baltimore Journal of Environmental Law* 1995; 5: 226–61.

70. Clark, D. L. *A Citizen's Guide to the World Bank Inspection Panel* (Washington, DC: Center for International Environmental Law; October, 1999).

71. Udall, L. *World Bank Inspection Panel* (Contributing Paper 126 to the World Commission on Dams, 2000).

72. Carrascott, E. R. and A. K. Guernsey. "The World Bank's Inspection Panel: Promoting True Accountability Through Arbitration," *Cornell International Law Journal* 2008; 41: 577–629.

73. Ibid.

74. Ibid.

75. Ibid.

76. Hunter, D. "Using the World Bank Inspection Panel to Defend the Interests of Project-Affected People," *Chicago Journal of International Law* 2003; 4: 201–11.

77. Clark, D. "Understanding the World Bank Inspection Panel," in D. Clark, J. Fox, and K. Treakle (Eds) *Demanding Accountability: Civil Society Claims and the World Bank Inspection Panel* (New York: Rowman and Littlefield; 2003, pp. 1–24).

78. Dunkerton, K. J. "The World Bank Inspection Panel and its Affect on Lending Accountability to Citizens of Borrowing Nations," *University of Baltimore Journal of Environmental Law* 1995; 5: 226–61.

79. Carrascott, E. R. and A. K. Guernsey. "The World Bank's Inspection Panel: Promoting True Accountability Through Arbitration," *Cornell International Law Journal* 2008; 41: 577–629.

80. Clark, D. "Understanding the World Bank Inspection Panel," in D. Clark, J. Fox and K. Treakle (Eds) *Demanding Accountability: Civil Society Claims and the World Bank Inspection Panel* (New York: Rowman and Littlefield,; 2003, pp. 1–24).

81. Ibid.

82. Ibid.

83. Ibid.

84. World Bank Inspection Panel, *The Inspection Panel: Annual Report July 1 2010 to June 20, 2011* (Washington, DC: International Bank for Reconstruction and Development/World Bank Group; 2011).

85. International Bank for Reconstruction and Development and International Development Association (2009).

86. NTPC. "Environmental Study of Singrauli Area" (performed by Electricité de France [EdF] International, July 1991).
87. Clark, D. "Singrauli: An Unfulfilled Struggle for Justice," in D. Clark, J. Fox and K. Treakle (Eds) *Demanding Accountability: Civil Society Claims and the World Bank Inspection Panel* (New York: Rowman and Littlefield; 2003, pp. 167–90).
88. Clark, D. "Understanding the World Bank Inspection Panel," in D. Clark, J. Fox and K. Treakle (Eds) *Demanding Accountability: Civil Society Claims and the World Bank Inspection Panel* (New York: Rowman and Littlefield,; 2003, pp. 1–24).
89. Werlin, H. H. "Helping Poor Countries: A Critique of the World Bank," *Orbis* 2003; Fall: 757–65.
90. Woods, N. "The Challenge of Good Governance for the IMF and the World Bank Themselves," *World Development* 2000; 28(5): 823–41.
91. Goldman, M. "Imperial Science, Imperial Nature: Environmental Knowledge for the World (Bank)," in S. Jasanoff and M. Long Martello (Eds) *Earthly Politics: Local and Global in Environmental Governance* (Cambridge: MIT Press; 2004, pp. 55–80); see also Goldman, M. *Imperial Nature: The World Bank and Struggles for Social Justice in the Age of Globalization* (New Haven, CT: Yale University Press; 2005).
92. Long Martello, M. and S. Jasanoff, "Globalization and Environmental Governance," in S. Jasanoff and M. Long Martello (Eds) *Earthly Politics: Local and Global in Environmental Governance* (Cambridge: MIT Press; 2004, pp. 1–29).
93. Carrascott, E. R. and A. K. Guernsey, "The World Bank's Inspection Panel: Promoting True Accountability Through Arbitration," *Cornell International Law Journal* 2008; 41: 577–629.
94. Nelson, P. J. "Transparency Mechanisms at the Multilateral Development Banks," *World Development* 2001; 29(11): 1835–47.
95. Hunter, D. "Using the World Bank Inspection Panel to Defend the Interests of Project-Affected People," *Chicago Journal of International Law* 2003; 4: 201–11.

5 Information and the Extractive Industries Transparency Initiative

1. Kolstad, I. and A. Wiig. "Is Transparency the Key to Reducing Corruption in Resource-Rich Countries?" *World Development* 2009; 37(3): 521–32.
2. Steiner, R. "Public Oversight of Extractive Industries in Developing Countries," *Paper Presented at the Revenue Watch Institute Public Oversight of Extractive Industries Conference. Baku, Azerbaijan, September, 2007.*
3. Karl, T. L. "Understanding the Resource Curse," in S. Tsalik and A. Schiffrin (Eds) *Covering Oil: A Reporter's Guide to Energy and Development* (New York: Open Society Institute; 2005, 21–7).
4. For lengthy explorations of this concept, see Humphreys, M., J. D. Sachs and J. E. Stiglitz (Eds). *Escaping the Resource Curse* (New York: Columbia University Press; 2007); Soares de Oliveira, R. *Oil and Politics in the Gulf of Guinea* (London: Hurst and Company; 2007); Sachs, J. D. and A. M. Warner. "Natural Resource Intensity and Economic Growth," in J. Mayer, B. Chambers and A. Farooq (Eds), *Development Policies in Natural Resource Economies* (Cheltenham: Edward Elgar; 1999, pp. 13–38); Auty, R. M. (Ed.) *Resource Abundance and Economic Development* (Oxford: Oxford University Press; 2001); see also Karl, T. L. *The Paradox of Plenty: Oil Booms and Petro-States* (Berkeley: University of California Press; 1997).

5. Stevens, P. "Resource Curse and Investment in Energy Industries," in C. Cleveland (Ed.) *Encyclopedia of Energy* 5 (London: Elsevier; 2004; pp. 451–9).

6. Karl, T. L. "State Building and Petro Revenues," in M. Garcelon, E. W. Walker, A. Patten-Wood and A. Radovich (Eds) *The Geopolitics of Oil, Gas, and Ecology in the Caucasus and Caspian Sea Basin* (Berkeley: Berkeley Institute of Slavic, East European, and Eurasian Studies; 1998, pp. 3–14).

7. De Soysa, I. "The Resource Curse: Are Civil Wars Driven by Rapacity or Paucity?" in M. Berdal and D. M. Malone (Eds) *Greed and Grievance: Economic Agendas in Civil Wars* (London: Lynne Reinner; 2000, pp. 113–35).

8. Silberfein, M. "The Geopolitics of Conflict and Diamonds in Sierra Leone," *Geopolitics* 2004; 9(1): 213–41.

9. Le Billon, P. "Geopolitical Economy of Resource Wars," *Geopolitics* 2004; 9(1): 1–28.

10. Humphreys, M., J. D. Sachs and J. E. Stiglitz. "What is the Problem with Natural Resource Wealth?" in M. Humphreys, J. D. Sachs and J. E. Stiglitz (Eds) *Escaping the Resource Curse* (New York: Columbia University Press; 2007, pp. 1–21).

11. Karl (2005).

12. Ross, M. L. "The Political Economy of the Resource Curse," *World Politics* 1999; 51(2): 297–322.

13. Humphreys *et al.* (2007).

14. Maassarani, T. F., Margo. T. Drakos and J. Pajkowska, "Extracting Corporate Responsibility: Towards a Human Rights Impact Assessment," *Cornell International Law Journal* 2007; Winter: 40.

15. Le Billon, P. "Geopolitical Economy of Resource Wars," *Geopolitics* 2004; 9(1): 1–28; see also O'Lear, S. "Resources and Conflict in the Caspian Sea," *Geopolitics* 2004; 9(1): 161–75.

16. Karl (2005).

17. Sachs, J. D. and A. M. Warner. "The Big Push, Natural Resource Booms and Growth," *Journal of Development Economics* 1999; 59: 43–76; Stiglitz, J. E. "Making Natural Resources Into a Blessing Rather Than a Curse," in S. Tsalik and A. Schiffrin (Eds) *Covering Oil: A Reporter's Guide to Energy and Development* (New York: Open Society Institute; 2005, pp. 13–20).

18. Goldwyn, D. L. (Ed.) Drilling Down: The Civil Society Guide to Extractive Industry Revenues and the EITI (Washington, DC: Revenue Watch; 2008).

19. Mikesell, R. "Explaining the Resource Curse, with Special Reference to Mineral-Exporting Countries," *Resources Policy* 1997; 23(4): 191–7.

20. Tsalik, S. "The Hazards of Petroleum Wealth," in R. Ebel (Ed.) *Caspian Oil Windfalls: Who Will Benefit?* (Washington, DC: Open Society Institute; 2003, pp. 1–15).

21. Corden, W. M. and J. P. Neary. "Booming Sector and Dutch Disease Economics," *The Economic Journal* 1982; 92: 825–48.

22. Idemudia, U. "The resource curse and the decentralization of oil revenue: the case of Nigeria," *Journal of Cleaner Production* 2012; 35: 183–93.

23. Karl, T. L. "Crude Calculations: OPEC Lessons for the Caspian Region," in R. Ebel and R. Menon (Eds) *Energy and Conflict in Central Asia and the Caucasus* (New York: Rowman & Littlefield; 2000, pp. 29–54).

24. Steiner, R. "Public Oversight of Extractive Industries in Developing Countries," *Paper Presented at the Revenue Watch Institute Public Oversight of Extractive Industries Conference*, Baku, Azerbaijan, September, 2007; see also Sovacool, B. K. "The Costs of Failure: A Preliminary Assessment of Major Energy Accidents, 1907 to 2007," *Energy Policy* 2008; 36(5): 1802–20.

25. Omorogbe (2006, p. 45).
26. Auty, R. M. "Natural Resources and Civil Strife: A Two-Stage Process," *Geopolitics* 2004; 9(1): 29–48; see also Auty, R. M. "Transition to Mid-Income Democracies or to Failed States?" In R. M. Auty and I. de Soysa (Eds) *Energy, Wealth, and Governance in the Caucasus and Central Asia: Lessons Not Learned* (London: Routledge; 2006, pp. 3–16).
27. Dunning, T. and L. Wirpsa. "Oil and the Political Economy of Conflict in Columbia and Beyond: A Linkages Approach," *Geopolitics* 2004; 9(1): 81–92.
28. Omorogbe (2006, p. 44).
29. Watts, M. "Resource Curse? Governmentality, Oil, and Power in the Niger Delta, Nigeria," *Geopolitics* 2004; 9(1): 50–69.
30. Idemudia, U. "The resource curse and the decentralization of oil revenue: the case of Nigeria," *Journal of Cleaner Production* 2012; 35: 183–93.
31. Baer, R. "The Fall of the House of Saud," *Atlantic Monthly* 2003; May: 34–48.
32. Karl (1998); Karl (2005, p. 23).
33. Brock, G. *Global Justice: A Cosmopolitan Account* (Oxford: Oxford University Press; 2009).
34. International Council on Human Rights Policy, *Climate Change and Human Rights: A Rough Guide* (Versoix, Switzerland; 2008).
35. Ibid.
36. Kolstad, I. and A. Wiig, "Is Transparency the Key to Reducing Corruption in Resource-Rich Countries?" *World Development* 2009; 37(3): 521–32.
37. Ibid.
38. Williams, A. "Shining a Light on the Resource Curse: An Empirical Analysis of the Relationship Between Natural Resources, Transparency, and Economic Growth," *World Development* 2011; 39(4): 490–505.
39. Al Faruque, A. "Transparency in Extractive Revenues in Developing Countries and Economies in Transition: a Review of Emerging Best Practices," *Journal of Energy and Natural Resources Law* 2006; 24: 66–103.
40. Extractive Industries Transparency Initiative. *EITI Rules, 2011, Including the Validation Guide* (Oslo, Norway: EITI Secretariat; November, 2011).
41. Global Witness, A crude awakening: The Role of the Oil and Banking Industries in Angola's Civil War and the Plunder of State Assets (London: Global Witness; 1999).
42. Eigen, P. "Fighting Corruption in a Global Economy: Transparency Initiatives in the Oil and Gas Industry," *Houston Journal of International Law* 2006–2007; 29: 327–54.
43. Ibid.
44. Extractive Industries Transparency Initiative. *EITI Rules, 2011, Including the Validation Guide* (Oslo, Norway: EITI Secretariat; November, 2011).
45. Eigen (2006–2007).
46. Extractive Industries Transparency Initiative. *EITI Rules, 2011, Including the Validation Guide* (Oslo, Norway: EITI Secretariat; November; 2011).
47. Friedman, A. "Operationalizing the Rio Principles: Using the success of the extractive transparency initiative to create a frame work for Rio Implementation," *University of Botswana Law Journal* 2001; 12: 73–86.
48. Extractive Industries Transparency Initiative. *EITI Rules, 2011, Including the Validation Guide* (Oslo, Norway: EITI Secretariat; November; 2011).
49. Extractive Industries Transparency Initiative. *Factsheet 2012* (Oslo, Norway: EITI Secretariat).

50. Extractive Industries Transparency Initiative. *EITI Rules, 2011, Including the Validation Guide* (Oslo, Norway: EITI Secretariat; November, 2011).

51. Extractive Industries Transparency Initiative. *Factsheet 2012* (Oslo, Norway: EITI Secretariat).

52. Friedman, A. "Operationalizing the Rio Principles: Using the success of the extractive transparency initiative to create a frame work for Rio Implementation," *University of Botswana Law Journal* 2001; 12: 73–86.

53. Extractive Industries Transparency Initiative. *Extracting Data: An Overview of EITI Reports Published 2005–2011* (Oslo, Norway: EITI Secretariat; 2011).

54. Al Faruque, A. "Transparency in Extractive Revenues in Developing Countries and Economies in Transition: a Review of Emerging Best Practices," *Journal of Energy and Natural Resources Law* 2006; 24: 66–103.

55. Extractive Industries Transparency Initiative. *Nigeria EITI: Making Transparency Count, Uncovering Billions* (Oslo, Norway: EITI Secretariat, January 20, 2012).

56. Ibid.

57. Friedman, A. "Operationalizing the Rio Principles: Using the success of the extractive transparency initiative to create a frame work for Rio Implementation," *University of Botswana Law Journal* 2001; 12: 73–86.

58. Extractive Industries Transparency Initiative. *Nigeria EITI: Making Transparency Count, Uncovering Billions* (Oslo, Norway: EITI Secretariat, January 20, 2012).

59. Goldwyn, D. L. (Ed.) *Drilling Down: The Civil Society Guide to Extractive Industry Revenues and the EITI* (Washington, DC: Revenue Watch; 2008).

60. EITI (2012).

61. Friedman, A. "Operationalizing the Rio Principles: Using the success of the extractive transparency initiative to create a frame work for Rio Implementation," *University of Botswana Law Journal* 2001; 12: 73–86.

62. Eigen, P. "Fighting Corruption in a Global Economy: Transparency Initiatives in the Oil and Gas Industry," *Houston Journal of International Law* 2006–2007; 29: 327–54.

63. Ibid.

64. Hess, D. "Combating Corruption through Corporate Transparency: Using EnforcementDiscretion to Improve Disclosure," *Minnesota Journal of International Law* 2012; 21: 42–74.

65. Eigen, P. "Fighting Corruption in a Global Economy: Transparency Initiatives in the Oil and Gas Industry," *Houston Journal of International Law* 2006–2007; 29: 327–54.

66. Al Faruque, A. "Transparency in Extractive Revenues in Developing Countries and Economies in Transition: a Review of Emerging Best Practices," *Journal of Energy and Natural Resources Law* 2006; 24: 66–103.

67. Eigen, P. "Fighting Corruption in a Global Economy: Transparency Initiatives in the Oil and Gas Industry," *Houston Journal of International Law* 2006–2007; 29: 327–54.

68. Ibid.

69. Gillies, A. and A. Heuty, "Does Transparency Work? The Challenges of Measurement and Effectiveness in Resource-Rich Countries," *Yale Journal of International Affairs* 2011; 6: 25–42.

70. Genasci, M. and S. Pray, "Extracting Accountability: The Implications of the Resource Curse for CSR Theory and Practice," *Yale Human Rights and Development Law Journal* 2008; 37: 50.

71. Goldwyn, D. L. (Ed.) *Drilling Down: The Civil Society Guide to Extractive Industry Revenues and the EITI* (Washington, DC: Revenue Watch; 2008).

72. Al Faruque, A. "Transparency in Extractive Revenues in Developing Countries and Economies in Transition: a Review of Emerging Best Practices," *Journal of Energy and Natural Resources Law* 2006; 24: 66–103.

73. Gillies, A. and A. Heuty, "Does Transparency Work? The Challenges of Measurement and Effectiveness in Resource-Rich Countries," *Yale Journal of International Affairs* 2011; 6: 25–42.

74. Al Faruque, A. "Transparency in Extractive Revenues in Developing Countries and Economies in Transition: a Review of Emerging Best Practices," *Journal of Energy and Natural Resources Law* 2006; 24: 66–103.

75. Ibid.

76. Shaxson, N. *Poisoned Wells: The dirty politics of African oil* (Basingstoke: Palgrave MacMillan; 2008).

77. Kolstad, I. and A. Wiig, "Is Transparency the Key to Reducing Corruption in Resource-Rich Countries?" *World Development* 2009; 37(3): 521–32.

78. Al Faruque, A. "Transparency in Extractive Revenues in Developing Countries and Economies in Transition: a Review of Emerging Best Practices," *Journal of Energy and Natural Resources Law* 2006; 24: 66–103.

79. Kardon, A. "Response to Matthew Genasci & Sarah Pray, Extracting Accountability: Implications of the Resource Curse for CSR Theory and Practice," *Yale Human Rights and Development Law Journal* 2008; 11: 59–67.

80. Otusanya, O. J. "The role of multinational companies in tax evasion and tax avoidance: The case of Nigeria," *Critical Perspectives on Accounting* 2011; 22: 316–32.

81. Kolstad, I. and A. Wiig, "Is Transparency the Key to Reducing Corruption in Resource-Rich Countries?" *World Development* 2009; 37(3): 521–32.

82. Collier, P. *Implications of Changed International Conditions for EITI* (Oxford University: EITI Secretariat, 2008).

83. Al Faruque, A. "Transparency in Extractive Revenues in Developing Countries and Economies in Transition: a Review of Emerging Best Practices," *Journal of Energy and Natural Resources Law* 2006; 24: 66–103.

84. Eigen, P. "Fighting Corruption in a Global Economy: Transparency Initiatives in the Oil and Gas Industry," *Houston Journal of International Law* 2006–2007; 29: 327–54.

85. Extractive Industries Transparency Initiative. *EITI Rules, 2011, Including the Validation Guide* (Oslo, Norway: EITI Secretariat; November, 2011).

86. Eigen, P. "Fighting Corruption in a Global Economy: Transparency Initiatives in the Oil and Gas Industry," *Houston Journal of International Law* 2006–2007; 29: 327–54.

87. Fung, A., M. Graham, D. Weil and E. Fagotto, *The Political Economy of Transparency: What Makes Disclosure Policies Effective?* (Harvard Transparency Policy Project; December 2004).

88. Eigen, P. "Fighting Corruption in a Global Economy: Transparency Initiatives in the Oil and Gas Industry," *Houston Journal of International Law* 2006–2007; 29: 327–54.

89. Kardon, A. "Response to Matthew Genasci & Sarah Pray, Extracting Accountability: Implications of the Resource Curse for CSR Theory and Practice," *Yale Human Rights and Development Law Journal* 2008; 11: 59–67.

90. Quoted in Conti-Brown, P. "Increasing the Capacity for Corruption?: Law and Development in the Burgeoning Petro-State of Sao Tome e Principe," *Berkeley Journal of African American Law and Policy* 2010; 12: 33–65.

91. Smith, S. M., D. D. Shepherd and P. T. Dorward, "Perspectives on community representation within the Extractive Industries Transparency Initiative: Experiences from south-east Madagascar," *Resources Policy* 2012; 37: 241–50.

92. Sourcewatch, "Greenwashing," 2012. Available at: http://www.sourcewatch.org/index.php/Greenwashing (accessed: February 1, 2013).
93. Goldwyn, D. L. (Ed.) *Drilling Down: The Civil Society Guide to Extractive Industry Revenues and the EITI* (Washington, DC: Revenue Watch; 2008).
94. Hilson, G. "Corporate Social Responsibility in the extractive industries: Experiences from developing countries," *Resources Policy* 2012; 37: 131–7.
95. Slack, K. "Mission impossible?: Adopting a CSR-based business model for extractive industries in developing countries," *Resources Policy* 2012; 37: 179–84.
96. Weszkalnys, G. "The Curse of Oil in the Gulf of Guinea: A View from Sao Tome and Principe," *African Affairs* 2009; 108: 679–89.
97. Quoted in McKibben, B. "Global Warming's Terrifying New Math," *Rolling Stone* 2012; July 19: 32–44.
98. Kolstad, I. and A. Wiig, "Is Transparency the Key to Reducing Corruption in Resource-Rich Countries?" *World Development* 2009; 37(3): 521–32.
99. Ibid.
100. Eigen, P. "Fighting Corruption in a Global Economy: Transparency Initiatives in the Oil and Gas Industry," *Houston Journal of International Law* 2006–2007; 29: 327–54.
101. NRC (National Research Council). *Hidden Costs of Energy: Unpriced Consequences of Energy Production and Use* (Washington, DC: The National Academies Press; 2009).

6 Prudence and São Tomé e Príncipe's Oil Revenue Management Law

1. Conceição, P., R. Fuentes and S. Levine, *Managing Natural Resources for Human Development in Low-Income Countries* (UNDP Regional Bureau for Africa, WP 2011–002, December 2011).
2. Ibid.
3. *Economist*, "Oilfield Services: The Unsung Masters of the Oil Industry," July 21, 2012, p. 53.
4. Brown, M. A. and B. K. Sovacool. *Climate Change and Global Energy Security: Technology and Policy Options* (Cambridge: MIT Press; 2011).
5. International Atomic Energy Agency, *Analysis Of Uranium Supply To 2050* (IAEA; 2001).
6. Ibid, p. 5.
7. Adapted from Hubbert, M. K., "Energy Resources of the Earth," *Scientific American* 1971; September: 61.
8. Source: Emirates Center for Strategic Studies and Research, *The Future of Oil as a Source of Energy* (ECSSR; 2003).
9. Humphreys, M. and M. E. Sandbu, *The Political Economy of Natural Resource Funds* (New York: Columbia University; 2007).
10. Hess, D. "Combating Corruption through Corporate Transparency: Using Enforcement Discretion to Improve Disclosure," *Minnesota Journal of International Law* 2012; 21: 42–74.
11. Watts, M. J. "Righteous Oil: Human Rights, The Oil Complex, and Corporate Social Responsibility," *Annual Review of Environment and Resources* 2005; 30: 373–407.
12. Maximus, C. "Islamic Republic corruption scandal: $11 billion in oil money missing," March 28, 2011, available at http://iranchannel.org/archives/962 (accessed: February 1, 2013).

13. Bryan, S. and B. Hofmann, *Transparency and Accountability in Africa's Extractive Industries: The Role of the Legislature* (Washington, DC: National Democratic Institute for International Affairs, 2007), pp. 36–7.

14. Bryan and Hofmann (2007).

15. Al Faruque, A. "Transparency in Extractive Revenues in Developing Countries and Economies in Transition: a Review of Emerging Best Practices," *Journal of Energy and Natural Resources Law* 2006; 24: 66–103.

16. Kardon, A. "Response to Matthew Genasci & Sarah Pray, Extracting Accountability: Implications of the Resource Curse for CSR Theory and Practice," *Yale Human Rights and Development Law Journal* 2008; 11: 59–67.

17. Brock, J. and T. Cocks, "Nigeria Oil Corruption Highlighted by Audits," Reuters News Service, March 8, 2012.

18. Idemudia, U. "The resource curse and the decentralization of oil revenue: the case of Nigeria," *Journal of Cleaner Production* 2012; 35: 183–93.

19. "Nigeria's Oil: A desperate need for reform," *The Economist*, October 20, 2012, p. 44.

20. Weszkalnys, G. "The Curse of Oil in the Gulf of Guinea: A View from Sao Tome and Principe," *African Affairs* 2009; 108: 679–89.

21. Christoff, J. A. "Observations on the Oil for Food Program," *Testimony Before the Committee on Foreign Relations, U.S. Senate* (Washington, DC: US GAO; April 7, 2004).

22. Johnston, M. "Why Do So Many Anti-Corruption Efforts Fail?" *New York University Annual Survey of American Law* 2011–2012; 27: 467–96.

23. Douglas, M. "Benefit-Cost Analysis, Future Generations and Energy Policy: A Survey of the Moral Issues," *Science, Technology, & Human Values* 1980; 5(31): 3–10.

24. Brundtland, G. H. *Our Common Future* (Oxford University Press, Oxford: World Commission on Environment and Development; 1987).

25. Conceição, P., R. Fuentes and S. Levine, *Managing Natural Resources for Human Development in Low-Income Countries* (UNDP Regional Bureau for Africa, WP 2011–002, December 2011).

26. Truman, E. M. "Sovereign Wealth Funds: New Challenges from a Changing Landscape," Testimony before the Subcommittee on Domestic and International Monetary Policy, Trade and Technology, Financial Services Committee, US House of Representatives, September 10, 2008.

27. Castilla, L. M. "Possible Future for a Peruvian 10 Billion SWF," July 13, 2012, available at http://www.swfinstitute.org/tag/copper/ (accessed: February 1, 2013).

28. Diplomatic Courier, "The Importance of Sovereign Wealth Funds to the World Economy," November 25, 2010.

29. Humphreys, M. and M. E. Sandbu, *The Political Economy of Natural Resource Funds* (New York: Columbia University; 2007).

30. Al Faruque, A. "Transparency in Extractive Revenues in Developing Countries and Economies in Transition: a Review of Emerging Best Practices," *Journal of Energy and Natural Resources Law* 2006; 24: 66–103.

31. Sandbu, M. E. "Natural Wealth Accounts: A Proposal for Alleviating the Natural Resource Curse," *World Development* 2006; 34(7): 1153–70.

32. Al Faruque, A. "Transparency in Extractive Revenues in Developing Countries and Economies in Transition: a Review of Emerging Best Practices," *Journal of Energy and Natural Resources Law* 2006; 24: 66–103.

33. Conti-Brown, P. "Increasing the Capacity for Corruption?: Law and Development in the Burgeoning Petro-State of Sao Tome e Principe," *Berkeley Journal of African American Law and Policy* 2010; 12: 33–65.

34. Bell, J. C. and T. M. Faria, "Sao Tome and Principe Enacts Oil Revenue Law, Sets New Transparency, Accountability, and Governance Standards," *Oil, Gas & Energy Law Intelligence* 2005; 3(1): 1–8.
35. Sandbu, M. E. "Natural Wealth Accounts: A Proposal for Alleviating the Natural Resource Curse," *World Development* 2006; 34(7): 1153–1170.
36. Jones, B. *São Tomé and Príncipe – Maximizing Oil Wealth for Equitable Growth and Sustainable Socio-Economic Development* (African Development Fund; 2012); Segura, A. "Management of Oil Wealth Under the Permanent Income Hypothesis: The Case of Sao Tome and Principe" (IMF Working Papers 06/183, International Monetary Fund; 2006).
37. Jones, B. *São Tomé and Príncipe – Maximizing Oil Wealth for Equitable Growth and Sustainable Socio-Economic Development* (African Development Fund; 2012); Conti-Brown, P. "Increasing the Capacity for Corruption?: Law and Development in the Burgeoning Petro-State of Sao Tome e Principe," *Berkeley Journal of African American Law and Policy* 2010; 12: 33–65; Financial Standards Foundation, *Country Brief: Sao Tome and Principe* (March 15, 2010); and Segura, A. "Management of Oil Wealth Under the Permanent Income Hypothesis: The Case of Sao Tome and Principe" (IMF Working Papers 06/183, International Monetary Fund; 2006).
38. See Jones, B. *São Tomé and Príncipe – Maximizing Oil Wealth for Equitable Growth and Sustainable Socio-Economic Development* (African Development Fund; 2012); Segura, A. "Management of Oil Wealth Under the Permanent Income Hypothesis: The Case of Sao Tome and Principe" (IMF Working Papers 06/183, International Monetary Fund; 2006).
39. Weszkalnys, G. "The Curse of Oil in the Gulf of Guinea: A View from Sao Tome and Principe," *African Affairs* 2009; 108: 679–89.
40. See Conti-Brown, P. "Increasing the Capacity for Corruption?: Law and Development in the Burgeoning Petro-State of Sao Tome e Principe," *Berkeley Journal of African American Law and Policy* 2010; 12: 33–65; as well as Soares de Oliveira, R. *Oil and Politics in the Gulf of Guinea* (London: Hurst and Company; 2007).
41. Conti-Brown, P. "Increasing the Capacity for Corruption?: Law and Development in the Burgeoning Petro-State of Sao Tome e Principe," *Berkeley Journal of African American Law and Policy* 2010; 12: 33–65.
42. Weszkalnys, G. "The Curse of Oil in the Gulf of Guinea: A View from Sao Tome and Principe," *African Affairs* 2009; 108: 679–89.
43. Personal correspondence with Jan Hartman, December 2012.
44. Frynas, J. G., G. Wood and R. M. S. Soares de Oliveira, "Business and politics in São Tomé e Príncipe," *African Affairs* 2003; 102: 51–80.
45. Conti-Brown, P. "Increasing the Capacity for Corruption?: Law and Development in the Burgeoning Petro-State of Sao Tome e Principe," *Berkeley Journal of African American Law and Policy* 2010; 12: 33–65.
46. Heuty, A. *Can Natural Resource Funds Address the Fiscal Challenges of Resource-Rich Developing Countries?* (Washington, DC: Revenue Watch Institute; 2010).
47. Bell, J. C. and T. M. Faria, "Sao Tome and Principe Enacts Oil Revenue Law, Sets New Transparency, Accountability, and Governance Standards," *Oil, Gas & Energy Law Intelligence* 2005; 3(1): 1–8.
48. Bell, J. C. and T. M. Faria, "Sao Tome and Principe Enacts Oil Revenue Law, Sets New Transparency, Accountability, and Governance Standards," *Oil, Gas & Energy Law Intelligence* 2005; 3(1): 1–8; Conti-Brown, P. "Increasing the Capacity for Corruption?: Law and Development in the Burgeoning Petro-State of Sao Tome e Principe," *Berkeley Journal of African American Law and Policy* 2010; 12: 33–65.

49. International Monetary Fund. *Democratic Republic of Sao Tome and Principe* (Washington, DC: IMF Country Report No. 12/216; August, 2012).

50. Segura, A. "Management of Oil Wealth Under the Permanent Income Hypothesis: The Case of Sao Tome and Principe" (IMF Working Papers 06/183, International Monetary Fund; 2006).

51. Frynas, J. G., G. Wood and R. M. S. Soares de Oliveira, "Business and politics in São Tomé e Príncipe," *African Affairs* 2003; 102: 51–80.

52. Jones, B. *São Tomé and Príncipe – Maximizing Oil Wealth for Equitable Growth and Sustainable Socio-Economic Development* (African Development Fund; 2012).

53. IMF. "Democratic Republic of Sao Tome and Principe, Joint IMF/World Bank Debt Sustainability Analysis," July 6, 2012.

54. Frynas, J. G., G. Wood and R. M. S. Soares de Oliveira, "Business and politics in São Tomé e Príncipe," *African Affairs* 2003; 102: 51–80.

55. Jones (2012).

56. IMF (2012).

57. IMF (2012).

58. Conti-Brown, P. "Increasing the Capacity for Corruption?: Law and Development in the Burgeoning Petro-State of Sao Tome e Principe," *Berkeley Journal of African American Law and Policy* 2010; 12: 33–65.

59. Jones, B. *São Tomé and Príncipe – Maximizing Oil Wealth for Equitable Growth and Sustainable Socio-Economic Development* (African Development Fund; 2012).

60. Ibid.

61. Truman, E. M. "Sovereign Wealth Funds: New Challenges from a Changing Landscape," Testimony before the Subcommittee on Domestic and International Monetary Policy, Trade and Technology, Financial Services Committee, US House of Representatives, September 10, 2008.

62. Personal communication with Jan Hartman, December 2012.

63. International Monetary Fund. Democratic Republic of Sao Tome and Principe (Washington, DC: IMF Country Report No. 12/216; August, 2012).

64. Personal correspondence with Jan Hartman, December 2012.

65. Conti-Brown, P. "Increasing the Capacity for Corruption?: Law and Development in the Burgeoning Petro-State of Sao Tome e Principe," *Berkeley Journal of African American Law and Policy* 2010; 12: 33–65.

66. International Monetary Fund, "Democratic Republic of Sao Tome and Principe" (Washington, DC: IMF Country Report No. 12/216, August, 2012).

67. Heuty, A. *Can Natural Resource Funds Address the Fiscal Challenges of Resource-Rich Developing Countries?* (Washington, DC: Revenue Watch Institute; 2010).

68. Humphreys, M. and M. E. Sandbu, *The Political Economy of Natural Resource Funds* (New York: Columbia University; 2007).

69. Fasano, U. "Review of the Experience with Oil Stabilization and Savings Funds in Selected Countries," IMF Working Paper (Washington, DC: International Monetary Fund; 2000).

70. Davis, J., R. Ossowski, J. Daniel and S. Barnett. "Stabilization and Savings Funds for Nonrenewable Resources," in J. Davis, R. Ossowski and O. Fedelino (Eds) *Fiscal Policy Formulation and Implementation in Oil-Producing Countries*, (Washington, DC: International Monetary Fund; 2003).

71. That is partly because nobody wants these unproven blocks. There were open bidding auctions, yet nobody bid. So STP resorted to giving contracts to anybody who would agree to drill.

72. Murdock, J. *Governance of Natural Resources in São Tomé and Príncipe: A Case Study onOversight and Transparency of Oil Revenues* (International Alert;, November, 2009).

73. Financial Standards Foundation, *Country Brief: Sao Tome and Principe* (March 15, 2010).
74. Weszkalnys, G. "The Curse of Oil in the Gulf of Guinea: A View from Sao Tome and Principe," *African Affairs* 2009; 108: 679–89.
75. Al Faruque, A. "Transparency in Extractive Revenues in Developing Countries and Economies in Transition: a Review of Emerging Best Practices," *Journal of Energy and Natural Resources Law* 2006; 24: 66–103.

7 Intergenerational Equity and Solar Energy in Bangladesh

1. Rawls, J. "Justice as Fairness," in W. Sellars and J. Hospers (Eds) *Readings in Ethical Theory* (Englewood Cliffs: Prentice-Hall; 1970, pp. 578–95).
2. Taebi, B. *Nuclear Power and Justice between Generations: A Moral Analysis of Fuel Cycles* (The Netherlands: Center for Ethics and Technology; 2010).
3. Nussbaum, M. C. *Frontiers of Justice: Disability, Nationality, Species Membership* (Cambridge, MA: Belknap Press; 2006, pp. 9–10).
4. Rawls, J. *A Theory of Justice: Revised Edition* (Cambridge, MA: Belknap Press; 1999, p. 12).
5. See Sen, A. *Resources, Values and Development* (Oxford: Basil Blackwell; 1984); Sen, A. "Capability and well-being," in M. Nussbaum and A. Sen (Eds), *The Quality of Life* (Oxford: Oxford University Press; 1993, pp. 30–53); Sen, A. *Development as Freedom* (Oxford: Oxford University Press; 1999). See also Walsh, V. "Amartya Sen on Rationality and Freedom," *Science & Society* 2007; 71(1): 59–83; Page, E. A. "Intergenerational Justice of What: Welfare, Resources or Capabilities?," *Environmental Politics* 2007; 16(3): 453–69; Muraca, B. "Towards a fair degrowth-society: Justice and the right to a 'good life' beyond growth," *Futures* (2012, in press).
6. Corbridge, S. "Development as freedom: the spaces of Amartya Sen," *Progress in Development Studies* 2002; 2(3): 183–217.
7. Nussbaum, M. C. *Creating Capabilities: The Human Development Approach* (Cambridge: Belknap Press; 2011).
8. Nussbaum (2011, p. 18).
9. Page, E. A. "Intergenerational Justice of What: Welfare, Resources or Capabilities?," *Environmental Politics* 2007; 16(3): 453–69.
10. Nussbaum, M. C. *Frontiers of Justice: Disability, Nationality, Species Membership* (Cambridge, MA: Belknap Press; 2006, pp. 9–10).
11. Joshua 9:23 and 9:21.
12. International Energy Agency, United Nations Development Programme, United Nations Industrial Development Organization. *Energy Poverty: How to Make Modern Energy Access Universal?* (Paris: OECD; 2010).
13. Marmot Review Team. *The Health Impacts of Cold Homes and Fuel Poverty* (Friends of the Earth; May 2011).
14. United Nations Development Program. *Contribution of Energy Services to the Millennium Development Goals and to Poverty Alleviation in Latin America and the Caribbean* (Santiago, Chile: United Nations; October, 2009).
15. Barnes, D. F., K. Krutilla and W. Hyde. *The Urban Household Energy Transition: Energy, Poverty, and the Environment in the Developing World* (Washington, DC: Resources for the Future; 2004).
16. Myers, N. and J. Kent. "New Consumers: The Influence of Affluence on the Environment," *Proceedings of the National Academies of Science* 2003; 100(8): 4963–8.

17. Jacobson, A. and D. Kammen. "Letting the (energy) Gini out of the bottle: Lorenz curves of cumulative electricity consumption and Gini coefficients as metrics of energy distribution and equity," *Energy Policy* 2005; 33(14): 1825–32.

18. Rapanos Vassilis, T. and L. M. Polemis. "The Structure of Residential Energy Demand in Greece," *Energy Policy* 2006; 34: 31–7.

19. Rosas-Flores, J. A., D. Morillon Galvez, and J. L. Fernandez Zayas. "Inequality in the Distribution of Expense Allocated to the Main Energy Fuels for Mexican Households: 1968–2006," *Energy Economics* (in press, 2010).

20. Hunt, L. C., G. Judge and Y. Ninomiya. "Underlying Trends and Seasonality in UK Energy Demand: A Sectoral Analysis," *Energy Economics* 2003; 25: 93–118.

21. Fernandez, E., R. P. Saini and V. Devadas. "Relative Inequality in Energy Resource Consumption: A Case of Kanvashram Village, Pauri Garwal District, Uttranchall (India)," *Renewable Energy* 2005; 30: 763–72.

22. Druckman, A. and T. Jackson. "Measuring Resource Inequalities: The Concepts and Methodology for an Area-Based Gini Coefficient," *Ecological Economics* 2008; 65: 242–52.

23. Papathanasopoulou, E. and T. Jackson. "Measuring Fossil Resource Inequality- A longitudinal case study for the UK: 1968–2000," *Ecological Economics* (in press, 2010).

24. Tindale, S. and C. Hewett. "Must the Poor Pay More? Sustainable Development, Social Justice, and Environmental Taxation," in Andrew Dobson (Ed.) *Fairness and Futurity: Essays on Environmental Sustainability and Social Justice* (Oxford: Oxford University Press; 1999, pp. 233–48).

25. Tran Van Hoa. "Quality of Consumption: Some Australian Evidence," *Economics Letters* 1985; 19: 189–92.

26. Walker, G. and R. Day. "Fuel poverty as injustice: Integrating distribution, recognition and procedure in the struggle for affordable warmth," *Energy Policy* 2012; 49: 69–75.

27. Hussain, F. "Challenges and Opportunities for Investments in Rural Energy," presentation to the United Nations Economic and Social Commission for Asia and the Pacific (UNESCAP) and International Fund for Agricultural Development (IFAD) Inception Workshop on *Leveraging Pro-Poor Public-Private-Partnerships (5Ps) for Rural Development*, United Nations Convention Center, Bangkok, Thailand, September 26, 2011.

28. Masud, J., D. Sharan and B. N. Lohani, *Energy for All: Addressing the Energy, Environment, and Poverty Nexus in Asia* (Manila: Asian Development Bank; April, 2007).

29. Legros, G., I. Havet, N. Bruce *et al. The Energy Access Situation in Developing Countries: A Review Focusing on the Least Developed Countries and Sub-Saharan Africa* (New York: World Health Organization and United Nations Development Program; 2009).

30. World Health Organization. *Fuel for Life: Household Energy and Health* (Geneva: WHO; 2006, p. 8).

31. World Health Organization (2006, p. 8).

32. Masud *et al.* (2007).

33. Jin, Y. "Exposure to Indoor Air Pollution from Household Energy Use in Rural China: The Interactions of Technology, Behavior, and Knowledge in Health Risk Management," *Social Science & Medicine* 2006; 62: 3161–76.

34. Holdren, J. P. and K. R. Smith. "Energy, the Environment, and Health," in T. Kjellstrom, D. Streets and X. Wang (Eds) *World Energy Assessment: Energy and the Challenge of Sustainability* (New York: United Nations Development Programme; 2000, pp. 61–110).

35. United Nations Environment Programme 2000. *Natural Selection: Evolving Choices for Renewable Energy Technology and Policy* (New York: United Nations). Figures have been updated to $2010.

36. Legros *et al.* (2009).

37. International Energy Agency. *World Energy Outlook 2006* (Paris: OECD; 2006).

38. International Energy Agency, United Nations Development Programme, United Nations Industrial Development Organization (2010, p. 7).

39. Gaye, A. "Access to Energy and Human Development," *Human Development Report 2007/2008*. United Nations Development Program Human Development Report Office Occasional Paper (2007).

40. Gaye, A. (2007).

41. Jin *et al.* (2006).

42. Masud *et al.* (2007).

43. United Nations Development Program. *Energy after Rio: Prospects and Challenges* (New York: UNDP; 1997).

44. Sangeeta, K. "Energy Access and Its Implication for Women: A case study of Himachal Pradesh, India," presentation to the *31st IAEE International Conference Pre-Conference Workshop on Clean Cooking Fuels*, Istanbul, June 16–17, 2008.

45. Reddy, B. S., P. Balachandra and H. S. K. Nathan. "Universalization of Access to Modern Energy Services in Indian Households – Economic and Policy Analysis," *Energy Policy* 2009; 37: 4645–57.

46. United Nations Development Programme (1997).

47. Murphy, J. "Making the Energy Transition in Rural East Africa: Is Leapfrogging an Alternative?" *Technological Forecasting & Social Change* 2001; 68: 173–93.

48. Schaefer, M. "Water technologies and the environment: Ramping up by scaling down," *Technology in Society* 2008; 30: 415–22.

49. Sovacool, B. K. and I. M. Drupady. *Energy Access, Poverty, and Development: The Governance of Small-Scale Renewable Energy in Developing Asia* (New York: Ashgate; 2012).

50. Asaduzzaman, M., D. F. Barnes and S. R. Khandker. *Restoring the Balance: Bangladesh's Rural Energy Realities* (Washington, DC: World Bank Working Paper No. 181; 2010).

51. Chowdhury, B. H. *Survey of Socio-Economic Monitoring & Impact Evaluation of Rural Electrification and Renewable Energy Program* (Dhaka: Rural Electrification Board, June 24, 2010).

52. Total primary energy supply comprises the production of coal, crude oil, natural gas, nuclear fission, hydroelectric, and other renewable resources, plus imports, less exports, less international marine bunkers, and corrected for net changes in energy stocks.

53. Asaduzzaman, M., D. F. Barnes and S. R. Khandker. *Restoring the Balance: Bangladesh's Rural Energy Realities* (Washington, DC: World Bank Working Paper No. 181; 2010).

54. United Nations Development Program. *Energy and Poverty in Bangladesh: Challenges and the Way Forward* (Bangkok: UNDP Regional Center; 2007).

55. Kamal, A. and M. S. Islam. "Rural Electrification Through Renewable Energy: A Sustainable Model for Replication in South Asia," *Presentation to the SAARC Energy Centre, August 7–9, 2010, Hotel Sheraton, Dhaka, Bangladesh.*

56. Islam, S. "IDCOL SHS Program – A Sustainable Model for Rural Lighting." *Presentation to the Promoting Rural Entrepreneurship for Enhancing Access To Clean Lighting Conference, Delhi, October 28,* 2010a.

57. United Nations Development Program. *Overcoming Vulnerability to Rising Oil Prices: Options for Asia and the Pacific* (Bangkok: UNDP Regional Center; 2007).
58. Wiser, R., G. Barbose, C. Peterman and N. Darghouth. *Tracking the Sun II: The Installed Cost of Photovoltaics in the U.S. from 1998 to 2008* (Berkeley: Lawrence Berkeley National Laboratory; October 2009, LBNL-2674E).
59. M. Z. Rahman, "Multitude of progress and unmediated problems of solar PV in Bangladesh," *Renewable and Sustainable Energy Reviews* 2012; 16: 466–73.
60. Ahammed, F. and D. A. Taufiq, "Applications of Solar PV On Rural Development in Bangladesh," *Journal of Rural and Community Development* 2008; 3: 93–103.
61. Islam (2010a). –
62. This quote comes from one of 48 research interviews conducted in Bangladesh at 19 institutions and communities in five locations over the course of June 2009 to October 2010. It is presented anonymously to respect the wishes of the respondent and to adhere to institutional review board guidelines concerning human subjects research.
63. Islam (2010a).
64. Islam, S. *IDCOL Solar Home Systems Program: An Offgrid Solution in Bangladesh* (Dhaka: IDCOL; July 26, 2010b).
65. This quote comes from research interviews, see note 62.
66. Urmee, T and D. Harries. "Determinants of the success and sustainability of Bangladesh's SHS program," *Renewable Energy* 2011; 36 (11): 2822–30.
67. This quote comes from research interviews, see note 62.
68. Ibid.
69. Ibid.
70. Ibid.
71. Ibid.
72. Ibid.
73. Ibid.
74. Ibid.
75. Ibid.
76. Ibid.
77. Ibid.
78. Ibid.
79. Ibid.
80. Asaduzzaman, M., D. F. Barnes and S. R. Khandker. *Restoring the Balance: Bangladesh's Rural Energy Realities* (Washington, DC: World Bank Working Paper No. 181; 2010).
81. Barnes, D. F., S. R. Khandker and H. A. Samad. *Energy Access, Efficiency, and Poverty: How Many Households are Energy Poor in Bangladesh?* (Washington, DC: World Bank Development Research Group, June, Working Paper 5332; 2010).
82. This quote comes from research interviews, see note 62.
83. Ibid.
84. Chowdhury, B. H. *Survey of Socio-Economic Monitoring & Impact Evaluation of Rural Electrification and Renewable Energy Program* (Dhaka: Rural Electrification Board, June 24, 2010).
85. Source: Chowdhury (2010).
86. S. Chakrabarty and T. Islam, "Financial viability and eco-efficiency of the solar home systems (SHS) in Bangladesh," *Energy* 2011; 36: 4821–7.
87. Urmee, T. and D. Harries. "Determinants of the success and sustainability of Bangladesh's SHS program," *Renewable Energy* 2011; 36(11): 2822–30.

88. Kaneko, S. K. S., R. M. Shrestha and P. P. Ghosh. "Non-income factors behind the purchase decisions of solar home systems in rural Bangladesh," *Energy for Sustainable Development* 2011; 15 (2011): 284–92.
89. Ibid.
90. Ibid.
91. Ibid.
92. Ibid.
93. Ibid.
94. Ibid.
95. Ibid.
96. Mondal, Md. A. H. and A. K. M. Sadrul Islam. "Potential and viability of grid-connected solar PV system in Bangladesh," *Renewable Energy* 2011; 36: 1869–74.
97. Ibid.
98. Rahman, M. Z. "Multitude of progress and unmediated problems of solar PV in Bangladesh," *Renewable and Sustainable Energy Reviews* 2012; 16: 466–73.
99. This quote comes from research interviews, see note 62.
100. Schumacher, E. F. *Small is Beautiful: Economics as if People Mattered* (New York, NY, USA: Harper & Row; 1973).

8   Intragenerational Equity and Climate Change Adaptation

1. This chapter draws from original research from two previous studies: Sovacool, B. K., A. L. D'Agostino, H. Meenawat and A. Rawlani. "Expert Views of Climate Change Adaptation in Least Developed Asia," *Journal of Environmental Management* 2012; 97(30): 78–88; and Sovacool, B. K., A. D'Agostino, A. Rawlani and H. Meenawat. "Improving Climate Change Adaptation in Least Developed Asia," *Environmental Science & Policy* 2012; 21(8): 112–25.
2. For a survey of these arguments, see Adger, W. N. and S. Nicholson-Cole, "Ethical Dimensions of Adapting to Climate Change-Imposed Risks," in D. G. Arnold (Ed.) *The Ethics of Global Climate Change* (Cambridge: Cambridge University Press, 2011), pp. 255–71; Vanderheiden, S., *Atmospheric Justice: A Political Theory of Climate Change* (New York: Oxford University Press; 2008); Posner, E. A. and C. R. Sunstein, "Climate Change Justice," *Georgetown Law Journal* 2008; 96: 1565–612; Adger, W. N., J. Paavola and S. Huq, "Toward Justice in Adaptation to Climate Change," in W. N. Adger, J. Paavola, S. Huq and M. J. Mace (Eds) *Fairness in Adaptation to Climate Change* (Cambridge: MIT Press; 2006, pp. 1–19); Caney, S. "Cosmopolitan Justice, Rights and Global Climate Change," *Canadian Journal of Law and Jurisprudence* 2006; 19(2): 255–78; Caney, S. *Justice Beyond Borders: A Global Political Theory* (Oxford: Oxford University Press; 2005); Agarwal, A., S. Narain and A. Sharma, "The Global Commons and Environmental Justice – Climate Change," in J. Byrne, L. Glover and C. Martinez (Eds) *Environmental Justice: International Discourses in Political Economy – Energy and Environmental Policy* (New Jersey: Transaction Publishers; 2002, p. 173).
3. Walker, G. *Environmental Justice: Concepts, Evidence, and Politics* (London: Routledge; 2012, p. 179).
4. Adger, W. N., J. Paavola, S. Huq and M. J. Mace. "Preface," *Fairness in Adaptation to Climate Change* (Cambridge: MIT Press; 2006, p. xi).
5. Steger, T. *Making the Case for Environmental Justice in Central and Eastern Europe* (Budapest: CEU Center for Environmental Law and Policy; March, 2007).

6. Archer, D. *The Long Thaw* (Princeton: Princeton University Press; 2009).

7. Once emitted, a ton of carbon dioxide takes a very long time to process through the atmosphere. According to the latest estimates, one-fourth of all fossil fuel-derived carbon dioxide emissions will remain in the atmosphere for several centuries, and complete removal could take as long as 30,000 to 35,000 years. See Hansen, J., M. Sato, P. Kharecha, *et al.* "Target Atmospheric CO2: Where Should Humanity Aim?" *Atmospheric Science Journal* 2008; 2: 217–31; and Archer, D. "Fate of Fossil Fuel CO2 in Geologic Time," *Journal of Geophysical Research* 2005; 110: 26–31.

8. Victor, D., G. Morgan, J. Steinbruner and K. Ricke. "The geoengineering option: A last resort against global warming?" *Foreign Affairs*, 2009; 88: 65.

9. See Shue, H. "Responsibility to Future Generations and the Technological Transition," in W. Sinnott-Armstrong and R. B. Howarth (Eds) *Perspectives on Climate Change: Science, Economics, Politics, Ethics* (Amsterdam: Elsevier; 2005, pp. 265–83); and Shue, H. "Climate," in D. Jamieson (Ed.) *A Companion to Environmental Philosophy* (London and New York: John Wiley and Sons; 2001, pp. 450–77).

10. Nolt, J. "Greenhouse Gas Emissions and the Domination of Posterity," in D. G. Arnold (Ed.) *The Ethics of Global Climate Change* (Cambridge: Cambridge University Press; 2011, pp. 61–76).

11. See Shue, H. *Basic Rights: Subsistence, Affluence and U.S. Foreign Policy* (Princeton, NJ, USA: Princeton University Press; 1980); Shue, H. "Subsistence Emissions and luxury emissions," *Law Policy* 1993; 15: 39–59. See also Shue, H. "Human Rights, Climate Change, and the Trillionth Ton," in D. G. Arnold (Ed.) *The Ethics of Global Climate Change* (Cambridge: Cambridge University Press; 2011, pp. 292–314); and Rao, N. and P. Baer, " 'Decent Living' Emissions: A Conceptual Framework," *Sustainability* 2012; 4: 656–81.

12. Shue (1980).

13. Harris, P. G. "Introduction: Cosmopolitanism and Climate Change Policy," in P. G. Harris (Ed.) *Ethics and Global Environmental Policy: Cosmopolitan Conceptions of Climate Change* (Cheltenham, UK: Edward Elgar; 2011, pp. 1–19).

14. H. Shue. *Subsistence, Affluence and U.S. Foreign Policy*, 2nd edition (Princeton: Princeton University Press; 1996); Shue, H. "Global Environment and International Inequality," in S. M. Gardiner, S. Caney, D. Jamieson and H. Shue (Eds) *Climate Ethics: Essential Readings* (Oxford: Oxford University Press; 2010, pp. 101–11).

15. Carrell, S. "Ocean Acidification Rates Accelerating," *The Hindu* (December 11, 2009): 11.

16. Good, P., J. Caesar, D. Bernie *et al.* "A review of recent developments in climate change science. Part I: Understanding of future change in the large-scale climate system," *Progress in Physical Geography* 2011; 35(3): 281–96; Gosling, S. N., R. Warren, N. W. Arnell *et al.* "A review of recent developments in climate change science. Part II: The global-scale impacts of climate change," *Progress in Physical Geography* 2011; 35(4): 443–64.

17. Lamirande, H. R. "From Sea to Carbon Cesspool: Preventing the World's Marine Ecosystems from Falling Victim to Ocean Acidification," *Suffolk Transnational Law Review* 2011; 34: 183–217.

18. Walsh, M. W. and N. Schwartz, "Estimate of Economic Losses Now Up to $50 Billion," *New York Times*, November 1, 2012.

19. Reddy, B. S. and G. B. Assenza, "The Great Climate Debate," *Energy Policy* 2009; 37: 2997–3008.

20. Catarious, D. M. and R. E. Espach. *Impacts of Climate Change on Columbia's National and Regional Energy Security* (Washington, DC: CNA; 2009).
21. De La Torre, A., P. Fajnzybler and J. Nash. *Low-Carbon Development: Latin American Responses to Climate Change* (Washington, DC: World Bank Group; 2010).
22. Sovacool, B. K. "Conceptualizing Hard and Soft Paths for Climate Change Adaptation," *Climate Policy* 2011; 11(4): 1177–83; Sovacool, B. K. "Perceptions of Climate Change Risks and Resilient Island Planning in the Maldives," *Mitigation and Adaptation of Strategies for Global Change* (in press, 2012); Sovacool, B. K. "Expert Views of Climate Change Adaptation in the Maldives," *Climatic Change* (in press, 2012).
23. Meenawat, H. and B. K. Sovacool. "Improving Adaptive Capacity and Resilience in Bhutan," *Mitigation and Adaptation Strategies for Global Change* 2011; 16(5): 515–33.
24. Brown, M. A. and B. K. Sovacool. *Climate Change and Global Energy Security: Technology and Policy Options* (Cambridge: MIT Press; 2011).
25. Gordon, R. "Climate Change and the Poorest Nations: Further Reflections on Global Inequality," *University of Colorado Law Review* 2007; 78: 1559–1624.
26. CNA. *Climate Change, State Resilience, and Global Security Conference* (CNA Conference Center, Alexandria, Virginia, November 4, 2009).
27. Haines, A., K. R Smith, D. Anderson, P. R. Epstein, A. J. McMichael, I. Roberts, P. Wilkinson, J. Woodcock and J. Woods. "Policies for accelerating access to clean energy, improving health, advancing development, and mitigating climate change," *Lancet* 2007; 370: 1264–81.
28. Prouty, A. E. "The Clean Development Mechanisms and its Implications for Climate Justice," *Columbia Journal of Environmental Law* 2009; 34(2): 513–40.
29. Economics of Climate Adaptation Working Group. *Shaping Climate-Resilient Development: A Framework for Decision-Making* (New York: Climate Works Foundation; 2009).
30. CNA. *Climate Change, State Resilience, and Global Security Conference* (CNA Conference Center, Alexandria, Virginia, November 4, 2009).
31. Royal United Services Institute (RUSI). *Socioeconomic and Security Implications of Climate Change in China* (Washington, DC: CNA; November 4, 2009).
32. Meenawat, H. and B. K. Sovacool. "Adapting to Climate Change in Laos: Challenges and Opportunities," *TERI Information Digest on Energy and Environment* 2011; 10(4): 497–504.
33. Meenawat, H. and B. K. Sovacool. "Improving Adaptive Capacity and Resilience in Bhutan," *Mitigation and Adaptation Strategies for Global Change* 2011; 16(5): 515–33.
34. Rawlani, A. and B. K. Sovacool. "Building Responsiveness to Climate Change through Community Based Adaptation in Bangladesh," *Mitigation and Adaptation Strategies for Global Change* 2011; 16(8): 845–863.
35. World Health Organization. *Climate Change and Human Health* (Geneva: WHO; December, 2003).
36. Royal United Services Institute (RUSI). *Socioeconomic and Security Implications of Climate Change in China* (Washington, DC: CNA; November 4, 2009).
37. Sovacool, B. K. "Conceptualizing Hard and Soft Paths for Climate Change Adaptation," *Climate Policy* 2011; 11(4): 1177–83; Sovacool, B. K. "Perceptions of Climate Change Risks and Resilient Island Planning in the Maldives," *Mitigation and Adaptation of Strategies for Global Change* (in press, 2012); Sovacool, B. K.

"Expert Views of Climate Change Adaptation in the Maldives," *Climatic Change* (in press, 2012).

38. Rawlani, A. and B. K. Sovacool. "Building Responsiveness to Climate Change through Community Based Adaptation in Bangladesh," *Mitigation and Adaptation Strategies for Global Change* 2011; 16(8): 845–863.

39. CNA. *National Security and the Threat of Climate Change* (Alexandria: CNA Corporation; 2007).

40. Biermann, F. and I. Boas. "Protecting Climate Refugees: The Case for a Global Protocol," *Environment* 2008; 50(6): 8–16.

41. Quoted in Penn State, "Ethics in Climate Change," 2010, available at http://www.psu.edu/dept/liberalarts/sites/rockethics/climate/ (accessed: March 20, 2013).

42. Biermann, F. and I. Boas. "Protecting Climate Refugees: The Case for a Global Protocol," *Environment* 2008; 50(6): 8–16.

43. Kakissis, J. "Environmental Refugees Unable to Return Home," *New York Times*, January 23, 2010, p. 23.

44. Weir, K. "Don't Cry for Kiribati, Tuvalu, Marshall Islands, parts of Papua New Guinea, the Caribbean, Bangladesh, Africa...," *Pacific Ecologist* 2008; Winter: 2.

45. Smith, A. "Climate Refugees in Maldives Buy Land," *Tree Hugger* Press Release, November 16, 2008.

46. Ayers, J. and T. Forsyth. "Community-based adaptation to climate change: strengthening resilience through development," *Environment* 2009; 51(4): 22–31.

47. IPCC. *Climate Change 2007: Impacts, Adaptation, and Vulnerability. Contribution of Working Group II to the Fourth Assessment Report of the Intergovernmental Panel on Climate Change*, M. L. Parry, O. F. Canziani, J. P. Palutikof, P. J. Van der Linden and C. E. Hanson (Eds) (Cambridge: Cambridge University Press; 2007).

48. Boyle, M. and Dowlatabadi, H. "Anticipatory Adaptation in Marginalized Communities Within Developed Countries," in J. D. Ford and L. Berrang-Ford (Eds), *Climate Change Adaptation in Developed Nations: From Theory to Practice* (London: Springer; 2011) *Advances in Global Change Research* 2011; 42: 461–73.

49. Folke, C. "Resilience: The emergence of a perspective for social-ecological systems analyses," *Global Environmental Change* 2006; 16(3): 253–67.

50. Jerneck, A. and L. Olsson. "Adaptation and the Poor: Development, Resilience, and Transition," *Climate Policy* 2008; 8: 170–82.

51. Economics of Climate Adaptation Working Group. *Shaping Climate-Resilient Development: A Framework for Decision-Making* (New York: Climate Works Foundation; 2009).

52. GEF, LDCF resources now amount to more than half a billion dollars (US $537 Million) (2012).

53. Rockefeller Foundation. *Building Climate Change Resilience* (New York: Rockefeller Foundation White Paper; 2010).

54. Sovacool, B. K. "Sound Climate, Energy, and Transport Policy for a Carbon Constrained World," *Policy & Society* 2009; 27(4): 273–83.

55. Grasso, M. *Justice in Funding Adaptation under the International Climate Change Regime* (Springer; 2010).

56. GEF, *Accessing Resources Under the Least Developed Countries Fund* (Washington, DC: GEF; 2009).

57. Personal correspondence with the GEF, December 2012.

58. Asian Development Bank, Coastal Greenbelt Project (Loan 1353-BAN[SF]) in the People's Republic of Bangladesh (ADB; October, 2005).

59. Government of Bangladesh and United Nations Development Program. *Project Document: Community Based Adaptation through Coastal Afforestation in Bangladesh* (UNDP, Dec 10, 2008, Dhaka Office, Project ID, PIMS 3873); Government of Bangladesh. *Bangladesh Climate Change Strategy and Action Plan* (Dhaka: Ministry of Environment and Forest (MOEF); 2009).

60. Dorji, W., S. R. Bajracharya, K. Kunzang, D. R. Gurung, Joshi, S. R. and Mool, P. K. *Inventory of Glaciers and Glacial Lakes and Glacial Lake Outburst Floods Monitoring and Early Warning Systems in the Hindu Kush-Himalayan Region Bhutan* (ICIMOD, UNEP; 2001); National Environment Commission, Royal Government of Bhutan. *Bhutan National Adaptation Program of Action* (2008).

61. Cambodian Ministry of Environment. *National Adaptation Programme of Action to Climate Change (NAPA)* (Phnom Penh: Ministry of Environment; 2006); see also Ponlok, T. "Climate Change in Cambodia: What Does it Mean to Cambodia & National-Based Adaptation to Climate Change," presented at *Climate Change Repercussions in Rural Cambodia*, November 19, 2009, Phnom Penh.

62. Khan, T., D. Quadir, T. S. Murty, A. Kabir, F. Aktar and M. Sarker. "Relative Sea Level Changes in Maldives and Vulnerability of Land Due to Abnormal Coastal Inundation," *Marine Geodesy* 2002; 25: 133–43.

63. Matthew, R. A. 2007. "Climate Change and Human Security," in J. F. C. DiMento and P. Doughman (Eds) *Climate Change: What It Means for Us, Our Children, and Our Grandchildren* (Cambridge: MIT Press; 2007, pp. 161–80).

64. Asian Development Bank. *The Economics of Climate Change in Southeast Asia: A Regional Review* (Manila: ADB; April, 2009).

65. Australian Aid. *Australian Multilateral Assessment of the Least Developed Countries Fund (LDCF)* (Canberra, Australia; March, 2012).

66. Tol, R. "The marginal damage costs of carbon dioxide emissions: an assessment of the uncertainties," *Energy Policy* 2005; 33(16): 2064–74; Tol, R. S. J. "The Economics Effects of Climate Change," *Journal of Economic Perspectives* 2009; 23(2): 29–51.

67. Kopp, R. E. and B. K. Mignone. "The U.S. Government's Social Cost of Carbon Estimates after their First Year: Pathways for Improvement," *Economics* 2011–16, June 8, 2011.

68. Ackerman, F. and E. Stanton. "Climate Risks and Carbon Prices: Revising the Social Cost of Carbon," *Economics* 2012–10 | April 4, 2012. http://www.e3network.org/social_cost_carbon.html (accessed March 20, 2013).

69. Australian Aid. *Australian Multilateral Assessment of the Least Developed Countries Fund (LDCF)* (March 2012).

70. GEF. *Accessing Resources Under the Least Developed Countries Fund* (Washington, DC: GEF; 2009).

71. Personal correspondence with the GEF, December 2012.

72. Flam, K. H. and J. B. Skjaerseth. "Does adequate financing exist for adaptation in developing countries?" *Climate Policy* 2009; 9: 109–14; GEF. *Evaluation of the GEF Strategic Priority for Adaptation* (July 2011).

73. Flam, K. H. and J. B. Skjaerseth. "Does adequate financing exist for adaptation in developing countries?" *Climate Policy* 2009; 9: 109–14.

74. Füssel, H.-M. S. Hallegatte and M. Reder. "International Adaptation Funding," in O. Edenhofer, J. Wallacher, H. Lotze-Campen *et al.* (Eds) *Climate Change, Justice and Sustainability: Linking Climate and Development Policy* (Springer; 2012, pp. 311–30).

75. Flam, K. H. and J. B. Skjaerseth. "Does adequate financing exist for adaptation in developing countries?" *Climate Policy* 2009; 9: 109–14.

76. Donner, S. D., M. Kandlikar and H. Zerriffi. "Preparing to Manage Climate Change Financing," *Science* 2011; 334: 908–9.
77. Evaluation Department, Ministry of Foreign Affairs of Denmark. *Evaluation of the operation of the Least Developed Countries Fund for adaptation to climate change* (September 2009).
78. Grasso, M. *Justice in Funding Adaptation under the International Climate Change Regime* (Springer; 2010)
79. UNDP. *Evaluation of UNDP Work with Least Developed Countries Fund and Special Climate Change Fund Resources* (New York: UNDP Evaluation Office; July 2009).
80. Ibid.
81. Ibid.
82. Danish Ministry of Foreign Affairs. *Review of the follow up on the LDCF Evaluation and information update on the LDCF and SCCF* (GEF/LDCF.SCCF.9/Inf.7, October 20, 2010).
83. Climate Change Forum. *The Least Developed Countries Fund & the Special Climate Change Fund* (Factsheet, 2010).
84. Australian Aid. *Australian Multilateral Assessment of the Least Developed Countries Fund (LDCF)* (March 2012).
85. United Nations Framework Convention on Climate Change. *Report of the Global Environment Facility to the Conference of the Parties* (September 20, 2012), available at http://unfccc.int/resource/docs/2012/cop18/eng/06.pdf (accessed: February 1, 2013).
86. Ibid.
87. Ibid.
88. Nayar, A. "When the Ice Melts," *Nature* 2009; 461: 1042–6.
89. GEF. *Strategy on Adaptation to Climate Change for the Least Developed Countries Fund (LDCF) and The Special Climate Change Fund (SCCF)* (2009).
90. Personal correspondence with the GEF, December 2012.
91. Preston, B. L., R. M. Westaway and E. J. Yuen. "Climate adaptation planning in practice: an evaluation of adaptation plans from three developed nations," *Mitig Adapt Strateg Glob Change* 2011; 16: 407–38.
92. Mann, M. E. "Defining Dangerous Anthropogenic Interference," *PNAS* 2009; 106(11): 4065–6.
93. Antholis, W. and S. Talbott. *Fast Forward: Ethics and Politics in the Age of Global Warming* (Washington, DC: Brookings Institution Press; 2010).
94. Anderson, K. and A. Bows. "Beyond 'dangerous' climate change: emission scenarios for a new world," *Phil Trans R Soc A* (2011; 369: 20–44.
95. Projections assume IPCC's A1FI with growth allocations to countries based on International Energy Outlook (2008).
96. Projections assume IPCC's A1FI with growth allocations to countries based on International Energy Outlook (2008).
97. Republic of Maldives. *National Adaptation Program of Action: Republic of Maldives* (Malé: Ministry of Environment, Energy, and Water; 2007).
98. McGray, H., A. Hammill, R. Bradley, E. L. Schipper and J. E. Parry 2009. *Weathering the Storm: Options for Framing Adaptation and Development* (Washington, DC: World Resources Institute; 2009).
99. World Resources Institute (WRI) in collaboration with United Nations Development Programme, United Nations Environment Programme, and World Bank. *World Resources 2008: Roots of Resilience-Growing the Wealth of the Poor* (Washington, DC: WRI; 2008).

100. Klein, R. J. T, R. J.Nicholls and F. Thomalla. "Resilience to Natural Hazards: How Useful is This Concept?" *Environmental Hazards* 2003; 5: 35–45.
101. Evaluation Department, Ministry of Foreign Affairs of Denmark. *Evaluation of the operation of the Least Developed Countries Fund for adaptation to climate change* (September 2009).

9 Responsibility and Ecuador's Yasuní-ITT Initiative

1. Low, N. and B. Gleeson. "Environmental justice. Distributing environmental quality," chapter 5 in: N. Low and B. Gleeson, *Justice, society and nature: an exploration of political ecology* (London/New York: Routledge; 1998, pp. 102–32).
2. Caney, S. "Cosmopolitan Justice, Responsibility, and Global Climate Change," *Leiden Journal of International Law* 2005; 18: 747–75.
3. Steger, T. *Making the Case for Environmental Justice in Central and Eastern Europe* (Budapest: CEU Center for Environmental Law and Policy; March, 2007).
4. International Council on Human Rights Policy. *Climate Change and Human Rights: A Rough Guide* (Versoix, Switzerland; 2008).
5. Gardiner, S. M. "Ethics and Global Climate Change," in S. M. Gardiner, S. Caney, D. Jamieson and H. Shue (Eds) *Climate Ethics: Essential Readings* (Oxford: Oxford University Press; 2010, pp. 3–35).
6. Caney, S. "Cosmopolitan Justice, Responsibility, and Global Climate Change," *Leiden Journal of International Law* 2005; 18: 747–75; see also E. Neumayer, "In Defence of Historical Accountability for Greenhouse Gas Emissions", (2000) 33 Ecological Economics, 185–92.
7. Shue, H. *Subsistence, Affluence and U.S. Foreign Policy*, 2nd edition (Princeton: Princeton University Press; 1996); Shue, H. "Global Environment and International Inequality," in S. M. Gardiner, S. Caney, D. Jamieson and H. Shue (Eds) *Climate Ethics: Essential Readings* (Oxford: Oxford University Press; 2010, pp. 101–11).
8. MacFarquhar, N. "U.N. Deadlock on Addressing Climate Shift," *New York Times*, July 20, 2011.
9. Barry, B. "Justice as Reciprocity," in *Democracy, Power and Justice, Essays in Political Theory* (Oxford: Clarendon Press; 1989, pp. 463–94).
10. Barry, B. "Sustainability and Intergenerational Justice," *Theoria* 1997; 45(89): 43–65; Barry, B. (1999), "Sustainability and Intergenerational Justice," in A. Dobson (Ed.) *Fairness and Futurity: Essays on Environmental Sustainability and Social Justice* (New York: Oxford University Press; 1999, pp. 93–117).
11. Barry (1989, p. 403).
12. Thoreau, H. *Life and Writings* (Walden, MA, USA: Walden Woods Library; 2007).
13. Waskow, D. and C. Welch. "The Environmental, Social, and Human Rights Impacts of Oil Development," in S. Tsalik and A. Schiffrin (Eds) *Covering Oil: A Reporter's Guide to Energy and Development* (New York: Open Society Institute; 2005, 101–23).
14. Ibid.
15. Ibid.
16. Ibid.
17. Peterson, C. H., M. C. Kennicutt II, R. H. Green *et al.*, "Ecological Consequences of Environmental Perturbations Associated with Offshore Hydrocarbon Production: A Perspective on the Long-Term Exposures in the Gulf of Mexico," *Canadian Journal of Fisheries and Aquaculture Science* 1996; 53: 2637–54.

18. Silva, P. "National Energy Policy," Hearing Before the House Subcommittee on Energy and Air Quality (Washington, DC: Government Printing Office; February 18, 2001, pp. 113–16).

19. Waskow and Welch (2005).

20. Silva (2001).

21. Waskow and Welch (2005).

22. Sovacool, B. K. "Cursed by Crude: The Corporatist Resource Curse and the Baku-Tbilisi-Ceyhan (BTC) Pipeline," *Environmental Policy and Governance* 2011; 21(1): 42–57.

23. Andruchow, J. E., C. L. Soskolne, F. Racioppi *et al.* "Cancer Incidence and Mortality in the Industrial City of Sumgayit, Azerbaijan," *International Journal of Occupational Environmental Health* 2006; 12(3): 234–41; Bickham, J. W., C. W. Matson, A. Islamzadeh, *et al.* "The unknown environmental tragedy in Sumgayit, Azerbaijan," *Ecotoxicology* 2003; 12: 505–8.

24. Sovacool, B. K. "Abandoned Treaties, Environmental Damage, Fossil Fuel Dependence: The Coming Costs of Oil and Gas Exploration in the Arctic National Wildlife Refuge," *Environment, Development, and Sustainability* 2007; 9(2): 187–201; Sovacool, B. K. "Eroding Wilderness: The Ecological, Legal, Political, and Social Consequences to Oil and Natural Gas Exploration in the Arctic National Wildlife Refuge (ANWR)," *Energy & Environment* 2006; 17(4): 549–67.

25. United Nations Environment Program *et al.*, *Environment and Security: The Case of the Eastern Caspian Region* (UNEP, UNDP, UNECE, OSCE, REC and NATO; 2008).

26. Finer, M., C. N. Jenkins, S. L. Pimm, B. Keane and C. Ross, "Oil and Gas Projects in the Western Amazon: Threats to Wilderness, Biodiversity, and Indigenous Peoples," *PLOS One* 2008; 3(8): 1–9.

27. Watts, M. J. "Righteous Oil: Human Rights, The Oil Complex, and Corporate Social Responsibility," *Annual Review of Environment and Resources* 2005; 30: 373–407.

28. Sovacool, B. K. "A Transition to Plug-In Hybrid Electric Vehicles (PHEVs): Why Public Health Professionals Must Care," *Journal of Epidemiology and Community Health* 2010; 64(3): 185–87.

29. Godoy, J. "Auto Emissions Killing Thousands," *Common Dreams News Release,* June 3, 2004, available at http://www.countercurrents.org/en-godoy040604.htm (accessed March 20, 2013).

30. Ibid.

31. United Nations Development Program. *Yasuní ITT FAQs* (2011).

32. Larrea, C. "Ecuador's Yasuni-ITI Initiative: an option towards equity and sustainability," *The Road to Rio+20* (June, 2012, pp. 58–63).

33. Christian, M., M. Finer and C. Ross. "Last chance to save one of world's most species-rich regions," *Nature* 2008; 455: 861.

34. Davis, T. C. "Breaking Ground Without Lifting a Shovel: Ecuador's Plan to Leave Its Oil in the Ground," *Houston Journal of International Law* 2007–2008; 30: 243–58.

35. Finer, M., R. Moncel and C. N. Jenkins. "Leaving the Oil Under the Amazon: Ecuador's Yasuní-ITT Initiative," *Biotropica* 2010; 42(1): 63–6.

36. See Davis, T. C. "Breaking Ground Without Lifting a Shovel: Ecuador's Plan to Leave Its Oil in the Ground," *Houston Journal of International Law* 2007–2008; 30: 243–58; Larrea, C. "Ecuador's Yasuni-ITI Initiative: an option towards equity and sustainability," *The Road to Rio+20* (June, 2012, pp. 58–63).

37. Finer, M., V. Vijay, F. Ponce, C. N. Jenkins, and T. R. Kahn, "Ecuador's Yasuní Biosphere Reserve: a brief modern history and conservation challenges," *Environ Res Lett* 2009; 4: 1–9.

38. Martinez, E. "Leave the Oil in the Soil: The Yasuní Model," in K. Abramsky (Ed.) *Sparking a Worldwide Energy Revolution: Social Struggles in the Transition to a Post-Petrol World* (Oakland: AK Press; 2010, pp. 234–44).
39. Davis, T. C. "Breaking Ground Without Lifting a Shovel: Ecuador's Plan to Leave Its Oil in the Ground," *Houston Journal of International Law* 2007–2008; 30: 243–58.
40. Rival, L. "The Yasuní-ITT Initiative: Oil Development and Alternative Forms of Wealth Making in the Ecuadorian Amazon," QEH Working Paper Series – QEHWPS180, December, 2009.
41. Yasuní ITT Website, 2012, available at http://yasuni-itt.gob.ec/quees.aspx
42. United Nations Development Program, *Yasuni ITT FAQs* (2011).
43. Bass, M. S., M. Finer, C. N. Jenkins *et al.* "Global Conservation Significance of Ecuador's Yasuní National Park," *PLOS* 2010; 5(1): e8767.
44. Davis, T. C. "Breaking Ground Without Lifting a Shovel: Ecuador's Plan to Leave Its Oil in the Ground," *Houston Journal of International Law* 2007–2008; 30: 243–58.
45. Finer, M. *et al.* "Leaving the Oil Under the Amazon: Ecuador's Yasuní-ITT Initiative," *Biotropica* 2010; 42(1): 63–6.
46. Ibid.
47. United Nations Development Program. *Yasuni ITT FAQs* (2011).
48. Yasuní ITT, estado actual de la iniciativa yasuni ITT (Período del 10 de febrero al 15 de diciembre de 2011).
49. See Yasuní ITT Website, 2012, available at http://yasuni-itt.gob.ec/quees.aspx; Marx, E. "With \$116 Million Pledged, Ecuador Moves Forward With Plan to Protect Rainforest," *Science Insider*, January 13, 2012; United Nations Development Program. *Yasuni ITT FAQs* (2011).
50. Marx, E. "With \$116 Million Pledged, Ecuador Moves Forward With Plan to Protect Rainforest," *Science Insider*, January 13, 2012.
51. Finer, M., V. Vijay, F. Ponce, C. N. Jenkins, and T. R. Kahn, "Ecuador's Yasuní Biosphere Reserve: a brief modern history and conservation challenges," *Environ Res Lett* 2009; 4: 1–9.
52. Martinez, E. "Leave the Oil in the Soil: The Yasuni Model," in K. Abramsky (Ed.) *Sparking a Worldwide Energy Revolution: Social Struggles in the Transition to a Post-Petrol World* (Oakland: AK Press; 2010, pp. 234–44).
53. Davis, T. C. "Breaking Ground Without Lifting a Shovel: Ecuador's Plan to Leave Its Oil in the Ground," *Houston Journal of International Law* 2007–2008; 30: 243–58.
54. These statistics all come from Finer, M., V. Vijay, F. Ponce, C. N. Jenkins, and T. R. Kahn, "Ecuador's Yasuní Biosphere Reserve: a brief modern history and conservation challenges," *Environ Res Lett* 2009; 4: 1–9; Larrea, C. "Ecuador's Yasuni-ITI Initiative: an option towards equity and sustainability," *The Road to Rio+20* (June, 2012, pp. 58–63); Bass, M. S. *et al.*, "Global Conservation Significance of Ecuador's Yasuní National Park," *PLOS* 2010; 5(1): e8767; and Davis, T. C. "Breaking Ground Without Lifting a Shovel: Ecuador's Plan to Leave Its Oil in the Ground," *Houston Journal of International Law* 2007–2008; 30: 243–58.
55. Marx, E. "Conservation Biology: The Fight for Yasuní," *Science* 2010; 330: 1170–1.
56. Bass, M. S. *et al.*, "Global Conservation Significance of Ecuador's Yasuní National Park," *PLOS* 2010; 5(1): e8767.
57. These statistics come from Larrea, C. "Ecuador's Yasuni-ITI Initiative: an option towards equity and sustainability," *The Road to Rio+20* (June, 2012, pp. 58–63); Bass, M. S. *et al.*, "Global Conservation Significance of Ecuador's Yasuní National

Park," *PLOS* 2010; 5(1): e8767; and Marx, E. "Conservation Biology: The Fight for Yasuní," *Science* 2010; 330: 1170–1.

58. Finer, M., V. Vijay, F. Ponce, C. N. Jenkins, and T. R. Kahn, "Ecuador's Yasuní Biosphere Reserve: a brief modern history and conservation challenges," *Environ Res Lett* 2009; 4: 1–9.

59. Warnars, L. *The Yasuní -ITT Initiative: an international environmental equity mechanism?* (Master's thesis, Social and Political Sciences of the Environment, Radboud University, January 2010).

60. Ibid.

61. Bass, M. S. *et al.*, "Global Conservation Significance of Ecuador's Yasuní National Park," *PLOS* 2010; 5(1): e8767.

62. Martinez, E. "Leave the Oil in the Soil: The Yasuni Model," in K. Abramsky (Ed.) *Sparking a Worldwide Energy Revolution: Social Struggles in the Transition to a Post-Petrol World* (Oakland: AK Press; 2010, pp. 234–44).

63. See, among others, Davis, T. C. "Breaking Ground Without Lifting a Shovel: Ecuador's Plan to Leave Its Oil in the Ground," *Houston Journal of International Law* 2007–2008; 30: 243–58; Warnars, L. *The Yasuní -ITT Initiative: an international environmental equity mechanism?* (Master's thesis, Social and Political Sciences of the Environment, Radboud University, January 2010); Martinez, E. "Leave the Oil in the Soil: The Yasuni Model," in K. Abramsky (Ed.) *Sparking a Worldwide Energy Revolution: Social Struggles in the Transition to a Post-Petrol World* (Oakland: AK Press; 2010, pp. 234–44); Finer, M., V. Vijay, F. Ponce, C. N. Jenkins, and T. R. Kahn, "Ecuador's Yasuní Biosphere Reserve: a brief modern history and conservation challenges," *Environ Res Lett* 2009; 4: 1–9; Finer, M. *et al.* "Leaving the Oil Under the Amazon: Ecuador's Yasuní-ITT Initiative," *Biotropica* 2010; 42(1): 63–6.

64. Finer, M. *et al.* "Leaving the Oil Under the Amazon: Ecuador's Yasuní-ITT Initiative," *Biotropica* 2010; 42(1): 63–6.

65. Larrea, C. "Ecuador's Yasuni-ITI Initiative: an option towards equity and sustainability," *The Road to Rio+20* (June, 2012, pp. 58–63); Larrea, C. and L. Warnars, "Ecuador's Yasuní-ITT Initiative: Avoiding emissions by keeping petroleum underground," *Energy for Sustainable Development* 2009; 13: 219–23.

66. Martinez, E. "Leave the Oil in the Soil: The Yasuni Model," in K. Abramsky (Ed.) *Sparking a Worldwide Energy Revolution: Social Struggles in the Transition to a Post-Petrol World* (Oakland: AK Press; 2010, pp. 234–44).

67. Davis, T. C. "Breaking Ground Without Lifting a Shovel: Ecuador's Plan to Leave Its Oil in the Ground," *Houston Journal of International Law* 2007–2008; 30: 243–58.

68. Finer, M. *et al.* "Leaving the Oil Under the Amazon: Ecuador's Yasuní-ITT Initiative," *Biotropica* 2010; 42(1): 63–6; Larrea, C. "Ecuador's Yasuni-ITI Initiative: an option towards equity and sustainability," *The Road to Rio+20* (June, 2012, pp. 58–63).

69. Larrea, C. "Ecuador's Yasuni-ITI Initiative: an option towards equity and sustainability," *The Road to Rio+20* (June, 2012, pp. 58–63).

70. Larrea, C., N. Greene, L. Rival *et al. Yasuní-ITT: An Initiative to Change History* (United Nations Development Program Ecuador, GTZ and MDGIF; 2010).

71. Ibid.

72. Swing, K. "Fight for Yasuní Far from Finished," *Science* 2011; 331: 29.

73. Marx, E. "With $116 Million Pledged, Ecuador Moves Forward With Plan to Protect Rainforest," *Science Insider*, January 13, 2012.

74. Ibid.

75. Rival, L. "The Yasuní-ITT Initiative: Oil Development and Alternative Forms of Wealth Making in the Ecuadorian Amazon," QEH Working Paper Series – QEHWPS180, December, 2009.

76. Davis, T. C. "Breaking Ground Without Lifting a Shovel: Ecuador's Plan to Leave Its Oil in the Ground," *Houston Journal of International Law* 2007–2008; 30: 243–58.

77. United Nations Development Program, *Yasuní ITT FAQs* (2011).

78. Rival, L. "The Yasuní-ITT Initiative: Oil Development and Alternative Forms of Wealth Making in the Ecuadorian Amazon," QEH Working Paper Series – QEHWPS180, December, 2009.

79. Rival, L. "Ecuador's Yasuní-ITT Initiative: The old and new values of petroleum," *Ecological Economics* 2010; 70: 358–65.

80. United Nations Development Program, *Yasuní ITT FAQs* (2011).

81. Marx, E. "With $116 Million Pledged, Ecuador Moves Forward With Plan to Protect Rainforest," *Science Insider*, January 13, 2012.

82. Davis, T. C. "Breaking Ground Without Lifting a Shovel: Ecuador's Plan to Leave Its Oil in the Ground," *Houston Journal of International Law* 2007–2008; 30: 243–58.

83. Finer, M. *et al.* "Leaving the Oil Under the Amazon: Ecuador's Yasuní-ITT Initiative," *Biotropica* 2010; 42(1): 63–6.

84. United Nations Development Program, *Yasuní ITT FAQs* (2011).

85. Pelaez-Samaniego, M. R., M. Garcia-Perez, L. A. B. Cortez *et al.* "Energy sector in Ecuador: Current status," *Energy Policy* 2007; 35: 4177–89.

86. Rival, L. "The Yasuní-ITT Initiative: Oil Development and Alternative Forms of Wealth Making in the Ecuadorian Amazon," QEH Working Paper Series – QEHWPS180, December, 2009.

87. Amazon Watch. "Ecuador," 2012, available at http://amazonwatch.org/work/ecuador (accessed: February 1, 2013).

88. Bass, M. S. *et al.*, "Global Conservation Significance of Ecuador's Yasuní National Park," *PLOS* 2010; 5(1): e8767.

89. Marx, E. "Conservation Biology: The Fight for Yasuní," *Science* 2010; 330: 1170–1.

90. Finer, M. *et al.* "Leaving the Oil Under the Amazon: Ecuador's Yasuní-ITT Initiative," *Biotropica* 2010; 42(1): 63–6.

91. UN-REDD. "About the UN-REDD Program," 2012, available at http://www.un-redd.org/AboutUN-REDDProgramme/tabid/102613/Default.aspx (accessed: February 1, 2013).

92. Personal correspondence with Professor Ann Florini, December 2012.

93. Quoted in Rival, L. "Ecuador's Yasuní-ITT Initiative: The old and new values of petroleum," *Ecological Economics* 2010; 70: 358–65.

94. Rival, L. "Ecuador's Yasuní-ITT Initiative: The old and new values of petroleum," *Ecological Economics* 2010; 70: 358–65.

10 Conclusion – Conceptualizing Energy Justice

1. See Speth, J. G. "Time for Civic Unreasonableness," *Environment YALE: School of Forestry & Environmental Studies* 2008; Spring: 2–3; DiMento, J. F. C. and P. Doughman. "Making Climate Change Understandable," in J. F. C. DiMento and P. Doughman (Eds) *Climate Change: What It Means for Us, Our Children, and Our Grandchildren* (Cambridge: MIT Press; 2007, pp. 1–9); and Doulton, H. and K. Brown. "Ten years to prevent catastrophe? Discourses of climate change and

international development in the UK press," *Global Environmental Change* 2009; 19: 191–202.

2. Elkind, J. "Energy Security: Call for a Broader Agenda," in C. Pascual and J. Elkind (Eds) *Energy Security: Economics, Politics, Strategies, and Implications* (Washington, DC: Brookings Institution Press; 2010, pp. 119–48).

3. Wolfowitz, P. "Good Governance and Development – A Time for Action," *World Bank Press Release*, April 11, 2006.

4. Hawken, P. *The Ecology of Commerce: A Declaration of Sustainability* (New York: Harper Collins; 1994, p. 112).

5. Shue, H. "The Unavoidability of Justice," in *The International Politics of the Environment: Actors, Interests, and Institutions* (Oxford: Oxford University Press; 1992, pp. 373–97).

6. deShalit, A. *Why Posterity Matters: Environmental Policies and Future Generations* (London: Routledge; 1995, p. 6).

7. The results of this exercise are presented using slightly different categories than the principles identified here. For instance, rather than saying "intragenerational equity" respondents were asked to rate "climate change adaptation," and rather than rate "information" they were asked to rate "energy literacy." Also, categories not directly related to justice, but still involved in the survey, such as "water" or "decentralization," are not presented in the Table and Figure. For those who want to read more, full details of the survey are presented in Sovacool, B. K., S. V. Valentine, M. J. Bambawale, M. A. Brown, T. D. F. Cardoso, S. Nurbek, G. Suleimenova, L. Jinke, X. Yang, A. Jain, A. F. Alhajji and A. Zubiri. "Exploring Propositions about Perceptions of Energy Security: An International Survey," *Environmental Science & Policy* 2012; 16(1): 44–64; as well as Bambawale, M. J. and B. K. Sovacool. "Energy Security: Insights From A Ten Country Comparison," *Energy & Environment* 2012; 23(4): 559–86.

8. The concept here is sometimes called "lexical" ethical priorities, those that must take precedence over others. See Gordon, S. *Welfare, Justice, and Freedom* (New York: Columbia University Press; 1980); Kolm, S.-C. *Modern Theories of Justice* (Cambridge: MIT Press; 1996); and Walker, G. *Environmental Justice: Concepts, Evidence, and Politics* (London: Routledge; 2012).

9. The concept here is a "hierarchy of needs." See Maslow, A. *Motivation and personality*, 3rd rev. edition (New York: Harper & Row; 1954/1987).

10. The concept here is an "energy services ladder" with basic "rungs" of the latter taking priority over the other higher rungs. See Sovacool, B. K. "Conceptualizing Urban Household Energy Use: Climbing the 'Energy Services Ladder,' " *Energy Policy* 2011; 39(3): 1659–68; and Sovacool, B. K. "Security of Energy Services and Uses within Urban Households," *Current Opinion in Environmental Sustainability* 2011; 3(4): 218–24.

Index

CPSIA information can be obtained
at www.ICGtesting.com
Printed in the USA
FFOW04n0443271216
30774FF